BIM-enabled Cognitive Computing for Smart Built Environment
Potential, Requirements, and Implementation

Editor

Ibrahim Yitmen

Assoc. Professor in Management of Construction Production
Jönköping University, Gjuterigatan
Sweden

CRC Press
Taylor & Francis Group
Boca Raton   London   New York

CRC Press is an imprint of the
Taylor & Francis Group, an **informa** business

A SCIENCE PUBLISHERS BOOK

First edition published 2021
by CRC Press
6000 Broken Sound Parkway NW, Suite 300, Boca Raton, FL 33487-2742

and by CRC Press
2 Park Square, Milton Park, Abingdon, Oxon, OX14 4RN

Library of Congress Cataloging-in-Publication Data

Names: Yitmen, Ibrahim, editor.
Title: BIM-enabled cognitive computing for smart built environment : potential, requirements, and implementation / editor, Ibrahim Yitmen, Assoc. Professor in Management of Construction Production Jönköping University, Gjuterigatan, Sweden.
Description: First edition. | Boca Raton, FL : CRC Press, 2021. | Includes bibliographical references and index. | Contents: Overview of cyber-physical systems and enabling technologies in cognitive computing for smart built environment / Ibrahim Yitmen and Sepehr Alizadehsalehi -- Towards a digital twin-based smart built environment / Ibrahim Yitmen and Sepehr Alizadehsalehi -- BIM-IoT-integrated architectures as the backbone of cognitive buildings : current state and future directions / Ali Motamedi and Mehrzad Shahinmoghadam. | Summary: "This book explicitly brings together the BIM-enabled cognitive computing for smart built environment, and focuses on the potential, requirements and implementation of cognitive Internet of Things (CIoT) paradigm to buildings, Augmented Reality/Mixed Reality into cognitive building concepts, cognitive smart cities in its complexity, heterogeneity, and scope, and challenges of utilizing the big data generated by smart cities from a machine learning perspective. This book offers contributions to the processing, analysis, management, modeling, and simulation of big data and the associated applicability to BIM cognitive systems that will advance different aspects of future cognitive cities"-- Provided by publisher.
Identifiers: LCCN 2020051484 | ISBN 9780367861834 (hardcover)
Subjects: LCSH: Building information modeling.
Classification: LCC TH438.13 .B556 2021 | DDC 720.285--dc23
LC record available at https://lccn.loc.gov/2020051484

ISBN: 978-0-367-86183-4 (hbk)
ISBN: 978-0-367-71227-3 (pbk)
ISBN: 978-1-003-01754-7 (ebk)

Typeset in Palatino Roman
by Innovative Processors

Preface

The aim of this book is to provide knowledge into BIM-enabled cognitive computing for smart built environment involving cognitive network capabilities for smart buildings, integrating augmented reality/mixed reality into cognitive building concepts, Cognitive Internet of Things (CIoT) for smart cities, artificial intelligence applications for cognitive cities, and cognitive smart cities using Big Data and machine learning. It focuses on the potential, requirements and implementation of CIoT paradigm to buildings, artificial intelligence techniques, reasoning, and augmented reality/mixed reality into cognitive building concepts, the concept of cognitive smart cities in its complexity, heterogeneity, and scope, and the challenge of utilizing the Big Data generated by smart cities from a machine learning perspective.

This book comprises BIM-based and data-analytic research on cognitive IoT for smart buildings and cognitive cities using Big Data and machine learning as complex and dynamic systems. It presents applied theoretical contributions fostering a better understanding of such systems and the synergistic relationships between the motivating physical and informational settings. It also offers contributions relating to the ongoing development of BIM-based and data science technologies for the processing, analysis, management, modeling, and simulation of big and context data and the associated applicability to cognitive systems that will advance different aspects of future cognitive cities.

This book offers the required material to inform pertinent research communities about the state-of-the-art research and the latest development in the area of cognitive smart cities development, as well as a valuable reference for planners, designers, strategists, and ICT experts who are working towards the development and implementation of CIoT based on Big Data analytics and context-aware computing.

Ibrahim Yitmen

Acknowledgments

The completion of this book could not have been possible without the participation and assistance of the many contributors. The editor would like to thank everyone who contributed with their great insights and valuable contents for the book. This manuscript has resulted from the joint endeavors of 13 academics contributors drawn from all over the world. This is an accomplishment of collaboration and team spirit.

The editor wishes to convey his sincere thanks to all the individual chapter-authors for their dedication, dynamism, and excitement in co-producing this book.

Lastly, and truly important, I would like to thank my wife and son for their determined encouragement and support, which empowered me to undertake this book project.

Ibrahim Yitmen

About the Editor

Ibrahim Yitmen received his Ph.D. in architecture from Istanbul Technical University, Turkey. Dr Yitmen is currently working as an Associate Professor in Management of Construction Production at Jönköping University since February 2018. The focus of his research is mainly on innovation in construction involving socio-technical issues regarding digital transformation in the AEC industry. Recent special interest is on BIM-enabled cognitive computing for smart built environment, augmented reality/mixed reality for cognitive buildings, blockchain technology in construction supply chains, unmanned aerial vehicles and BIM integration for construction safety planning and monitoring, and automated BIM-based construction project progress monitoring. Dr Yitmen has published more than 70 research papers in referred international journals and in peer-reviewed conference proceedings. He is currently one of the guest editors of the *MDPI Journal of Applied Sciences* for the special issue on 'cognitive buildings'. Dr Yitmen has been serving on the scientific committees of international conferences held by the International Council for Research and Innovation in Building and Construction (CIB), American Society of Civil Engineers (ASCE), and European Council on Computing in Construction (EC³). Dr Yitmen is an active member of the American Society for Engineering Management, USA since 2010.

List of Contributors

Sepehr Alizadehsalehi completed his BSc in Civil Engineering, MSc in Construction Management & Technologies, MSc in Project Management, and Ph.D. in Construction Management & Innovation. He is an experienced construction manager with over seven years of experience in managing commercial and residential projects and more than eight years of experience in exploring and researching the most relevant and applicable construction management technologies, such as BIM, extended reality (XR), reality capture technologies, lean and collaborative working in construction, Digital Twins, scan-to-BIM, design management, safety management, and construction progress monitoring. Currently, he is continuously involved in R&D studies and technology development platforms for future applications of Digital Twins associated with artificial intelligence, Big Data analytics, and advanced computing at the McCormick School of Engineering of Northwestern University at Evanston, Illinois, USA.

Pardis Pishdad-Bozorgi is an Associate Professor at Georgia Institute of Technology. Her research and teaching efforts target two intertwined tracks, namely (1) integrating project delivery and trust-building and (2) integrating technologies for creating and operating a smart built-environment. She has over 40 peer-reviewed journals and conference publications in the emerging fields of Building Information Modeling (BIM), Internet of Things (IoT), blockchain, and Integrated Project Delivery (IPD). Dr Pishdad has been frequently acknowledged with numerous national and international awards for her excellence in research and teaching. She holds a Ph.D. degree in Environmental Design and Planning (Virginia Tech), three Master's degrees in the fields of Civil Engineering (Virginia Tech), Project Management (Harvard), and Architecture (University of Tehran), and a Bachelor's degree in Architectural Engineering (Azad University of Shiraz).

Stephan Embers studied Applied Computer Science at the Ruhr University Bochum and graduated in 2018 with a Master of Science degree. At the moment he is working as a Research Assistant at the Chair of Computing in Engineering at the Ruhr-University Bochum in Germany. There his field of research includes augmented and virtual reality in construction as well as human-machine interaction and construction safety.

Xinghua Gao is an Assistant Professor with the Myers-Lawson School of Construction at Virginia Tech. He received his Ph.D. degree in Building Construction and M.Sc. in Computational Science and Engineering from Georgia Institute of Technology, M.Sc. in Structural Engineering from Cardiff University, UK, and B.Eng. in Civil Engineering from Central South University, China. Prior to pursuing his Ph.D., he worked as a structural/Building Information Modeling (BIM) engineer at the China Institute of Building Standard Design & Research. His research interests lie in BIM and Internet of Things (IoT)-enabled smart built environments that encompass automated data collection, analysis, and visualization for more efficient, effective, and secured construction and facilities management.

Patrick Herbers studied Applied Computer Science at the Ruhr University Bochum and is now writing his Ph.D. thesis on indoor localization and augmented reality at the Chair of Computing in Engineering.

Markus König is Professor for Computing in Engineering (CIE) at the Ruhr-University Bochum, Germany. Previously, he was Assistant Professor of Theoretical Methods for Project Management at the Bauhaus University Weimar, Germany. He obtained his Ph.D. in Civil Engineering from the Leibniz-University, Hanover, Germany in 2003. His research interests include building information modeling, construction simulation and optimization, knowledge management in construction, intelligent computing in engineering, and computational steering. He was part of the expert panel which developed the 'Road Map for Digital Design and Construction' for the Federal Ministry of Transport and Digital Infrastructure (BMVI). Since 2016, he is one of the project leaders of the implementation team commissioned by the BMVI. He is currently head of the German BIM Competence Center for the Digitization of Construction. He has published more than 130 scientific journal and conference papers.

Ali Motamedi is Professor at *École de Technologie Supérieure* (ÉTS). He studied and conducted research at University of Tehran, Concordia University and Osaka University. His research investigates the application of sensor networks, ambient and artificial intelligence, and visual analytics for the design, construction, and management of built facilities. His interests also reside in the cyber-physical interactions provided by Digital Twins of facilities through an integration of digital models and the Internet of Things (IoT). Ali is the holder of several national and international scholarships, fellowships, and awards, including JSPS, FQRNT, NSERC, and the Merit Award. The results of his research have been extensively published and won multiple best-paper awards. He has also many years of consulting/design experience in the area of information technology and has participated in several large-scale IT projects.

Gozde Basak Ozturk is an Assistant Professor at the Faculty of Engineering, Department of Civil Engineering in Aydin Adnan Menderes University. She focuses on developing innovative integrated architecture/ engineering/construction/facility management. Her publications are on emerging fields, such as Building Information Modeling, organizational learning, knowledge management, collaborative design, performance measurement, 3D printing, and lean construction with many articles published in key journals and conferences. She holds a Ph.D. degree in Architecture branched in Construction Management (Izmir Institute of Technology), Master's degrees in Civil Engineering (European University of Lefke), Master's degrees in Business Administration (European University of Lefke), and Architecture (European University of Lefke).

Mehrzad Shahinmoghadam is a Ph.D. student at *École de Technologie Supérieure*, the Department of Construction Engineering. His research focuses on three major areas: (i) investigating best practices for creating Digital Twins of buildings by integrating disparate sources of lifecycle data within cloud environments, (ii) acquisition and reuse of the knowledge about dynamic behavior of buildings by applying data-driven and knowledge-based methods over the semantically-linked building databases, and (iii) maintain the human-in-the-loop paradigm for building lifecycle management by creating immersive and intuitive visual analytics interfaces.

Lavinia Chiara Tagliabue is an Assistant Professor at the University of Brescia, Department of Civil, Environmental, Architectural Engineering and Mathematics (DICATAM) and collaborator at the Department Information Engineering (DII). During her Ph.D. at Politecnico di Milano on Technology and Design for Environmental Quality at building and urban scale, she worked on the sustainability assessment of office buildings and energy requirements, energy-saving measures and renewable energy integration in the built environment (BiPV). She works on energy simulation and interoperability between Building Information Modeling and Building Energy Modeling to provide an uninterrupted chain of information between disciplines and support the design process of NZEB (Nearly Zero Energy Buildings). The research at the University of Brescia is focused on cognitive buildings, Digital Twins, and smart connections with a user-centered design approach (behavioral design) able to promote a new vision of the building as a service provider to enhance and improve the user experience.

Jong Han Yoon is a Ph.D. student in Building Construction at Georgia Institute of Technology. He has been a Graduate Research and Teaching Assistant at the School of Building Construction at Georgia Institute of Technology. He received his B.Eng. in Architectural Engineering, BBA in Business Administration, and M.Eng. in Architectural Engineering from Ajou University. Before pursuing his Ph.D., he worked as a research assistant at Ajou University Construction Engineering Management (AUCEM) laboratory. His research interests center on the blockchain-applied construction supply chain.

Mario Wolf is a Ph.D. researcher at the Chair of Digital Engineering. He is a trained technical assistant for business informatics and has studied applied informatics at the Ruhr University Bochum with focus on industrial and management informatics. He received his doctorate in 2018 at the Chair for IT in Mechanical Engineering under Prof. Abramovici on the topic of 'Smart Maintenance Assistance System for Process Engineering Plants in Industry 4.0', in which the reference architecture model Industry 4.0 was applied in an integrated use case. A fully dynamic, context-sensitive augmented reality application was made possible through seamless data management. His research focuses on user support in content generation for mixed reality applications and their use in technical training and engineering. He teaches object-oriented programming and software engineering.

Sven Zentgraf studied Applied Computer Science at the Ruhr University Bochum, and is now working as a research associate on BIM-related projects with focus on authoring and maintaining interconnected data at the Chair of Computing in Engineering.

List of Figures

List of Tables

Introduction

Cyber-Physical Systems (CPSs) are integrated monitoring, sensing and actuating systems that build a two-way connection between the physical space and cyber components to autonomously manage processes, information, and resources. Application of CPSs, such as Internet of Things (IoT), robotization, Digital Twin, and cognitive computing in the AEC industry will significantly revolutionize the way infrastructure and buildings are built, managed, and connected to other autonomous systems.

A cognitive system can sense, use signified knowledge, learn from its experience, collect knowledge, describe itself, receive instructions, be aware of its own behavior and capability, along with reacting in a manner that roughly simulates the process of cognition in the human mind. Cognitive computing is infusing and developing into scopes of machine learning (ML), deep learning (DL), natural language processing (NLP), computer vision (CV), and artificial intelligence (AI). The enormous amount of both structured and unstructured data created from IoT-based sensor devices, digital devices, and software applications is growing. Latest researches depict that a vast volume of data is created in Big Data. The cognitive system is shaped by numerous aspects, such as smart city platform, IoT layer, and data layer. The themes of user involvement and occupancy management grow in consequence, particularly in the latest era of cognitive buildings, by which benefits could be gained from self-learning and data analytics-based capabilities. Buildings are transforming from adaptive and predictive to cognitive assets, capable of interacting with human activity contained by them, across a significant scope of innovations. Along with the developments of building automation allocates designing and operating buildings informed of the conditions monitored contained by them, but also capable of responding to various inputs. When data is automatically handled, and buildings react consequently, buildings are called 'Cognitive buildings'. The Cognitive Internet of Things (CIoT) is identified as a network or an environment wherein everyone and everything is connected. Cognitive smart city implies the integration of evolving IoT and smart city technologies, their created Big Data, and artificial intelligence techniques. CIoT architecture-based smart city network involves key elements of CIoT architecture, sensor modules, machine learning, and semantic modeling. Smart buildings in the smart city platform deliver data gathered from sensors, including various human aspects, such as emotions, voice, brain activity, etc.

Building Information Modeling (BIM) currently confronts major challenges where controlling Big Data, IoT, and artificial intelligence are indicated as prospective solutions to automation and inclusion of broader environmental settings. The requirement to monitor and control assets (manufactured elements, buildings, bridges, etc.) during their lifecycle, combined with developments in technological facilities, have shifted numerous research topics into studying Digital Twin (DT) uses and potential. DT comprises sensors and measurement technologies, IoT, simulation, and modeling and machine learning technologies. IoT-cloud communication models and Big Data generated by IoT devices ensue enhanced potential and incremental data of cloud services to create a dynamic digital replica of buildings and city that integrates each sub-DT (e.g. building DT and bridge DT).

Creating DTs on blockchain can facilitate identification of and tracking them globally with high accuracy. Blockchain-compelling transparency and accountability through encryption and controlling mechanisms (e.g. smart contracts and chain code) can be utilized to preserve IoT systems and devices, such as temperature sensors, security cameras, air quality sensors, etc. from probable attacks, as it facilitates devices to make security decisions without depending on a central authority in smart city ecosystem.

Chapter 1 presents an overview of Cyber-Physical Systems (CPSs) and enabling technologies in cognitive computing for smart built environment, focusing on CPSs in AEC industry, Big Data with cognitive computing, cognitive building, IoT-enabled smart cities, and smart cities from a machine learning perspective.

Chapter 2 presents a comprehensive review of the state-of-the-art about Big Data with cognitive computing, Digital Twin in smart cities, BIM with Digital Twin, and blockchain with Digital Twin for smart built environment.

Chapter 3 sheds light on the definition of cognitive buildings and their key enablers and discusses the key requirements needed to create cognitive buildings, such as applying cognitive computing over BIM-IoT integrated architecture, and presents an overview of a high-level holistic architecture that can be used to create cognitive buildings.

Chapter 4 examines the IoT that can be integrated into BIM and determines the framework of the transformation of model data on Digital Twin for use throughout the project lifecycle.

Chapter 5 depicts how immersive technologies are used to create an augmented computer environment, how the data is integrated into the BIM, and the Digital Twin of the building which is a smart entity expected to manipulate decision-making process throughout the project lifecycle.

Chapter 6 shows various examples of smart maintenance use cases in civil engineering, featuring different methods, tools, and processes to augment the workers' experience and simplify the access to context-relevant data in cognitive buildings.

Chapter 7 explores the potential of blockchain in enhancing Cyber-Physical Security in smart built environments, highlighting the key aspects of Cyber-Physical Security in the smart built environment context, discussing the potential blockchain applications, and sharing directions for future research.

Chapter 8 presents a case study of eLUX lab cognitive building in the smart campus of the University of Brescia, involving a cognitive building approach by considering an energy management system that is able to regulate the HVAC system based on an occupancy rate model, including users' habits and IAQ, and provided in real-time by IoT sensors.

Ibrahim Yitmen

Contents

Overview of Cyber-Physical Systems and Enabling Technologies in Cognitive Computing for Smart Built Environment

Ibrahim Yitmen[1*] and Sepehr Alizadehsalehi[2]

[1] Department of Construction Engineering and Lighting Science, School of Engineering, Jönköping University, Gjuterigatan 5, 551 11, Jönköping, Sweden
[2] Project Management Program, Department of Civil and Environmental Engineering, Northwestern University, Evanston, IL, USA

1.1 Introduction

Cyber-Physical Systems (CPSs) comprising interconnected and integrated smart systems can transform the architecture, engineering, and construction (AEC) industry and contribute to the development of Construction 4.0. The built environment, being a significant constituent of the projected Construction 4.0, utilizes innovative CPS connecting smart buildings through IoT/enabled smart city network in real-time (Pardis *et al.*, 2020). Many researchers and industry professionals envision that CPS will augment the AEC industry through connected and autonomous systems to efficiently improve communication, operation, safety, and performance. CPS applications are predicted to significantly improve the approach in construction projects that are planned, managed, built, and connected to other autonomous systems. A smart built environment provides a fully integrated and networked connectivity between virtual assets and physical assets.

*Corresponding author: ibrahim.yitmen@ju.se

A clear understanding of the requirements, processes, and characteristics of CPS is essential to make for proper and precise CPS implementation in the AEC industry. Currently, there is limited literature about CPS technologies/processes in the AEC industry. For this purpose, in this chapter, a comprehensive review regarding CPS and its different levels of architecture, CPS in the AEC industry, enabling technologies in cognitive computing (CC), and smart cities and digital technologies are presented. Overall, this study provides a thorough exploration of using CPS to solve a variety of construction project management issues effectively and efficiently. More prominently, this study provides a roadmap for future efforts to implement CPS technologies in the AEC industry to contribute to the development of Construction 4.0 besides creating a smart built environment.

1.2 Cyber-Physical Systems (CPSs)

The key technological concept of Industry 4.0 is CPSs (Borangiu *et al.*, 2019). In the last couple of years, the importance of CPSs to optimizing industrial processes has led to a significant increase in sensitized production environments. Data collected in this context allows new intelligent solutions to support decision processes or to enable predictive actions. A CPS is a system with continuous automatic linking among the world of material and smart digital components that can sense, manage, and control the physical world (Klinc and Turk, 2019). CPSs are integrated systems that build a bi-directional link among the physical environment and cyber parts to manage methods, information, and supplies autonomously. CPSs are integrated multiple systems that create a bi-directional link among the physical environment and cyber components to manage techniques, methods, information, and supplies autonomously. They are working with different sensors, processors, and actuators, which can be monitored through computers provided with feedback loops with physical facilities.

CPSs run complex analytics through connectivity strength. Complex inference within a CPS occurs by a centralized analytic hub in which knowledge is acquired from raw data. Based on the knowledge inferred from the data and the help of the Internet of Things (IoT) control, commands are sent to the physical asset. CPS has extensive applications in medical operation, military systems, manufacturing systems, monitoring systems, traffic control and safety, and power generation. Jazdi (2014) stated that in contrast to the conventional embedded systems, created as separate devices, the emphasis of CPS of Industry 4.0 is on networking numerous devices. The IoT is a technology/system of interconnected computing and digital-based devices and machines with unique identifiers and data transferability over a network without requiring human-to-human or human-to-computer interaction. Recently, various industries have witnessed an increase in generated data in the form of structured and unstructured data from IoT-based digital devices, sensor tools, and software applications.

1.2.1   CPS 5C Level Architecture

Santos *et al.* (2017) describe CPSs as an augmentation of embedded systems linking the digital and physical worlds through integrating complex data processing from numerous interacted physical elements, such as people, machinery, equipment, sensors. Lee *et al.* (2015) stated that the main functional components of a CPS are advanced connectivity and intelligent data management, analytics, and computational capability. Based on this conceptual guideline, a 5C (Connection, Conversion, Computation, Cognition, and Configuration) architecture for application objectives was proposed (Muhuri *et al.*, 2019). Figure 1.1 illustrates the detailed CPS 5C architecture.

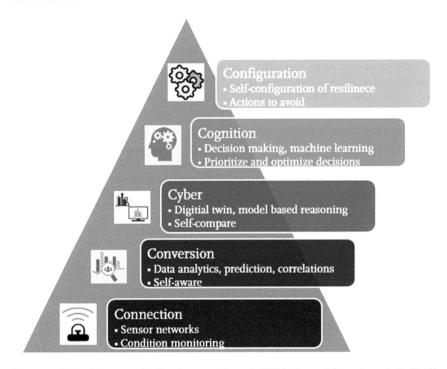

Fig. 1.1: 5C architecture for implementation of CPS (adapted from Lee *et al.*, 2015)

The proposed levels describe the architecture of a CPS from the preliminary data acquired over analytics to the ultimate value creation on five distinct levels (Lee *et al.*, 2015):

i. *The level of smart connections*: Gathering precise and reliable data from different components and sections is the primary phase in developing a CPS application, which might be measured directly by sensors or acquired from manufacturing systems. It comprises 'plug & play' devices, independent communications, and sensor networks. Data acquiring and transmission within selected devices utilizing uniform procedures have to be as forthright as possible for this level to operate under expectations.

ii. *The level of data-to-information conversion*: There exist services, tools, and procedures built on data stored at Level 1, that are being retrieved and utilized for predictions, correlations, statistics, and management to support decision making.

iii *The cyber level*: It performs as a central data hub within this pile. The primary model for this level is the Digital Twin (DT), digital representation of an object that exists or will exist in the physical world. When the data stored at the level of smart connections and analyzed at the data to information conversion level is located in the framework of the advanced data mining model, interconnections, simulations, and analytics turn out to be actual and practical.

iv *The cognition level*: Learning can proceed beyond and utilize artificial intelligence (AI) technologies to make progressive decisions, diagnostics, and machine learning (ML). It is presumed that a majority of information, which cannot be handled using conventional approaches, will require analyzing with machine learning as the scope and variety of data would be very big for the analysis to be processed manually by people writing algorithms.

v. *The configuration level*: It is the feedback from cyberspace to physical space, which functions as managerial control to make machines self-adaptive and self-configuring. This can lead to autonomous, smart, self-learned, and automatic configuration of cyber systems that can intelligently respond to environmental changes and user requirements.

Accordingly, the ever-growing use of sensors, reality capture technologies and networked machines has resulted in the continuous generation of high volume data, called Big Data. In such an environment, CPS, with all levels combined, will have a significant impact on the improvements to reach the goal of intelligent, resilient, and self-adaptable machines.

1.2.2 CPSs in AEC Industry

Various AEC researchers and industry professionals project that construction projects will be augmented by connected and autonomous systems to improve communication, operation, safety, and performance very shortly (Nunes *et al.*, 2015; Correa, 2018). Based on the current practices in the AEC industry, CPS could be seen as one of the best technologies to solve the recurrent problems faced in the industry. Its applications predicted to significantly ameliorate the approach construction projects are planned, managed, built, and connected to other autonomous systems (Linares *et al.*, 2019). A CPS is a bi-directional interaction among the physical components and cyber components of a system. In this process, sensors gather the physical components data. These data are then automatically transferred to cyber-components, and finally, the analyzed data are transmitted into required information through cyber processes. The necessary measures are decided and built on the earlier analysis and assessment and are transmitted back to physical components through actuators.

The built assets and construction sites are considered as the components of physical CPS, and Building Information Modeling (BIM) can be comprehended as a key CPS facilitator, demonstrating cyber counterparts and connecting them by their physical and other cyber components, comprising cloud servers and knowledge management centers (Linares *et al.*, 2019). However, this real-time interaction is exceptionally challenging because of interoperability among the CPSs and BIM platforms (Correa and Maciel, 2018).

Some of the current applications of bi-directional CPSs in the construction industry are intelligent building systems (Zheng, 2018), smart traffic management systems (Teizer *et al.*, 2010), and construction site safety operations (Teizer *et al.*, 2010). Most of these applications in the construction industry are unidirectional and much the same as IoT implementations, such as the automation of recurring and basic activities, enhanced productivity, modularization, safety, and the integration of future technology tendencies that improve project outcomes. Nevertheless, the more critical issues to consider for driving CPS in the industry involve a higher level of CPSs integration, standards definition, and the advancement of the supportive technologies for CPSs. CPSs are encouraged by numerous technologies that, based on their implementation level, can be separated into two major categories:

i. The various technologies, which are known, accepted, and implemented in construction projects like BIM, reality capturing technologies, portable devices, GPS, and UAVs, are currently utilized in different construction progressions and have the potential to encourage future higher levels of CPSs implementation.
ii. Trendy technologies that currently are not fully used. However, they can take CPSs applications to the next level, such as IoT, extended reality (XR) technologies (virtual reality (VR)/augmented reality (AR)/mixed reality (MR)), robotics, machine learning (ML), artificial intelligence (AI), etc.

As mentioned, there are various trendy technologies/systems for cyber applications in the AEC industry, such as IoT, fifth-generation (5G) wireless technology, XR, AI, DT, blockchain, robotization, cognitive computing (CC), and cloud computing (Chiarello *et al.*, 2018; Erboz, 2017; Santos *et al.*, 2017; Vaidya *et al.*, 2018):

Internet of Things (IoT): IoT technologies are facilitating continuous interoperability and improved connectivity between the physical and cyber world with potential benefits in different applications, including smart homes, smart buildings, smart cities, and others (Faheem *et al.*, 2018). The concept is to connect everything complex enough to be connected through the switch to the Internet and can be transmitted from the Internet. Therefore, the vital point is being equipped with smart sensors, which are sensing what is occurring around them in real-time. AEC industry has begun to use the IoT as products and services, including Information and Communications Technology (ICT) or semantic web technologies, to enhance the sharing and communication systems. The AEC industry with accessing and improving required standards for more reliable interoperability by other systems is currently ready for crucial changes that empower efficiency, well-being, process improvement, and new instruments (Rezgui *et al.*, 2011; Jacobsson *et al.*, 2017). Current applications include data sharing, digitalization of design and construction process, cost estimation, automation in buildings operations, increasing safety management, sustainability management, etc. (Abanda *et al.*, 2013; Woodhead *et al.*, 2018).

Fifth generation (5G) wireless technology: Increased bandwidth, speed, and improved latency widely will soon be provided by fifth generation (5G) wireless technology in all industries. It will allow AEC technologies to offer novel solutions to improve remote and co-located creative and interactive collaboration of project teams with multimodal information input (Alizadehsalehi *et al.*, 2020). This improvement by 5G is vital for different types of CPSs applications as this technology will enable the industry to deliver higher multi-Gbps peak data speeds, more reliability, ultra-low latency, massive network capacity, increased availability, and more consistent user experience to more stakeholders.

Artificial Intelligence (AI): Due to the necessity for synthetic thinking with non-human decision-makers and AI's ability to utilize advanced algorithms to 'learn' from 'big data' and then use the gained knowledge, it is predicted that AI will be vital for the implementation of autonomous CPS. Moreover, AI can tackle nonlinear and complicated practical problems and once trained, could undertake predictions and generalizations at a very high speed. Therefore, AI-based systems have the potential to be used for high-level systems, such as smart buildings, smart infrastructure, and even smart cities.

Extended Reality (XR): New applications of XR (VR/AR/MR) will influence project advancement, and they serve as an interface to visualize, manage, and help to access information/data from the cyber systems. XR provides a wide variety and a vast number of levels in the virtuality of partially sensor inputs to immersive virtuality. Applications of XR include safety management, design review, progress monitoring, remote control operations, etc. (Alizadehsalehi *et al.*, 2020).

Digital Twins (DT): The CPS being a principal component of Industry 4.0 (Linares *et al.*, 2019), requires the DT as the representation of the physical object for the development, analysis, and control of the production process (Uhlemann *et al.* 2017). A DT is a digital replica, a cyber-model of its physical opponent.

Blockchain: The framework of technology has the capability to resolve certain issues in CPSs implementation in the AEC industry by sustaining the practices related to data verification and enhanced cyber-security. Blockchain facilitates the AEC industry to authenticate shared data and hence sustain trust in information exchanges (Gries *et al.*, 2018), support supply chain management for the purpose of material control, and provide improvements in BIM databases (Lanko, Vatin *et al.*, 2018). Consequently, Blockchain can improve security and information sharing of CPS.

Robotization: It can transform digital information into the design (for example, by a mounting, adding – digital printing or removing) of the material world. The German Industry 4.0 Working Group (Group, 2013) called this new idea a 'smart factory', as it enables important improvements to the before-known industrial processes.

Cognitive Computing (CC): This covers a pile of emerging technologies, including Big Data, machine learning, cognitive algorithms, and AI. It attempts to simulate human thought processes in the computer model – using the DT. Conti *et al.* (2017) demonstrated that CC is replicating (in the cyber world) the method that users are examining and handling data in the physical world.

Cloud Computing: ICT infrastructure delivers warehousing, processing, and communication services as a service. Virtualization of hardware and networks enables it to effectively approach the suitable facilities, to ensure privacy, security, resilience, etc. Cloud computing performs as platforms facilitating Industry 4.0 partners' improved integration (Erboz, 2017), delivering a range of services to the smart factories of the future to integrate improved manufacturing and logistics processes (Marques *et al.*, 2017) and predictably provide a road to cloud manufacturing (Smit *et al.*, 2016).

The idea of 'smart city' is to deal with challenges faced through the growth of towns, that denotes "the effective integration of physical, digital, and human systems in the built environment to deliver a sustainable, prosperous, and inclusive future for its citizens" (ISO/IEC, 2015). Based on NIST (2019), the term 'smart' refers to the development, integration, and use of intelligent methods/systems based on ICT and CPS. CPS represents the smart systems that comprise physical engineered interacting networks, computational parts, and anything explaining related smart systems, such as the machine-to-machine (M2M), IoT, digital city. A CPS comprises two main parts, which all researchers in this area are trying to connect them to enable innovative applications and services (Gao *et al.*, 2019):

- The physical parts, which contain a device, a machine, a building, or an infrastructure.
- The digital/cyber parts which constitute the software system, communication network, and the data. This part represents the state of the physical part digitally and influences it by automated control or informing people of control actions.

Building systems already with limited system connectivity of IoT devices incorporate proprietary networks of sophisticated sensors and tools in energy systems, security systems, and emerging smart home devices. These smart building sensor networks show potential platforms for the deployment of further generalized IoT networks and are sources of occupancy and space, which are able to enhance new IoT systems significantly (Gries *et al.*, 2018; Gao *et al.*, 2019).

1.3 Enabling Technologies in Cognitive Computing (CC)

1.3.1 Cognitive Computing and Related Technologies

IBM created the cognitive computing (CC) term for the first time to explain systems that have the learning ability from a broad range of datasets, present reasons and interact with humans by natural languages, and gain their experiences in the context (Mohammadi and Al-Fuqaha, 2018). Cognitive science is the basis of studies related to cognition, like human cognition, which has three significant mechanisms, including perception, action, and learning. A cognitive-based system is able to reason, use expressed knowledge, learn from its own experiences, gather knowledge, define itself, understand its capabilities and behaviors, and respond like the human mind (Sheth, 2016). CC combines cognitive science and computer science to mimic human thinking systems in a computerized/

programmed environment to manage the human brain's real-time performance and mind-like function (Preissl *et al.*, 2012). Cognitive-based systems perceive objects the way humans do and identify unknown objects that they have not seen before being trained to recognize/identify as done in traditional ML processes (Schmidt, 2017).

CC has the ability to extract concepts, emotions, objects, keywords, and relationships from massive unstructured data, and use AI, computer vision (CV), natural language processing (NLP), and enhanced ML algorithms to process analytics, make predictions, and hypothesize in a scalable way to quickly learned/trained and improve as it discovers knowledge to forecast future events with purpose and estimated strength (Capuano and Toti, 2019). Xu *et al.* (2019) presented the Venn diagram, which shows the relationships between CC, AI, ML, DL, CV, and NLP. As shown in Fig. 1.2, DL is a subset of ML; CC, NLP, and CV are subsets of AI; and AI is a subset of CC.

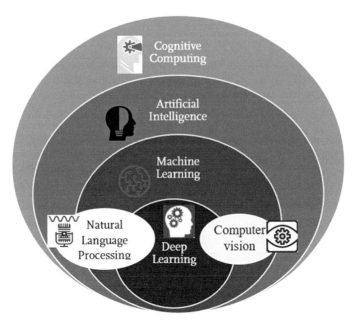

Fig. 1.2: The relations between CC's technologies (adapted from Xu *et al.*, 2019)

- AI refers to the simulation of human intelligence on to a machine to make the machine programmed and proficient at thinking like humans and mimicking their actions to identify and apply the right piece of 'knowledge' at a given step to solve a problem.
- ML is an application of AI, and it is a technology that provides a system that can automatically learn and improve itself from examples

and teach to solve problems without being programmed manually (Alpaydin, 2020).

- DL at the forefront of ML in AI has networks capable of learning unsupervised from unstructured or unlabeled data. Learning at DL can be supervised, semi-supervised, or unsupervised (Zhang *et al.*, 2019). Systems can be trained to behave with human brain-like thinking, using DL algorithms.
- NLP is a subfield of AI that deals with the interaction between computers and humans using the natural language. NLP deals with the interaction between computers and humans using the natural language and reads, deciphers, understands, and converts these human written or spoken languages into a formal representation easy for computers to manipulate (Young *et al.*, 2018).
- CC is an interdisciplinary scientific field that deals with transforming data from digital images or videos to high-level understandable data by computers. The ultimate goal in these technologies/sciences is automating the tasks that the human visual system can do.

The cognitive IoT (CIoT) is the extension of IoT, where IoT systems are equipped with CC approaches, allowing them to learn and reason over data and extract deep, actionable insights while building a network where the physical world and the digital world blend. The CIoT is recognized as an environment or a network that connects all devices (Zhang *et al.*, 2012), and it is considered in the existence of different sorts of structures based on prominent agents and architectures in cognitive systems. Through this mechanism, the cognition of IoT systems endows them with the freedom to operate intelligently and autonomously. Subsequently, the IoT systems not only learn and reason based on the experiences that they gain from their interactions with their counterparts and their environments in general, but their learning and reasoning abilities are improved at the cadence of the new information that they acquire. The concept of CIoT can integrate, improve performance, and achieve intelligence, and it is used to analyze the perceived information based on prior knowledge, make intelligent decisions, and perform adaptively and control actions. The CIoT architecture based on the smart city network, has three key components of CIoT architecture: sensor components, ML, and semantic modeling.

1.3.2   Big Data with CC

Big Data is a field that treats ways to analyze, systematically extract information from, or otherwise, deal with data sets that are too large or complex to be dealt with by traditional data-processing application software (Kambatla *et al.*, 2014; Breur, 2016). CC will bring a high level of fluidity to analytics. Based on Bedeley *et al.*'s (2018) research, the use of Big Data analytics consists of 5V categories, such as value, velocity, volume, veracity, and variety, while cognitive consists of four categories, such as observation, decision, interpretation, and evaluation. The cognitive system's Big Data is influenced by various factors, like smart city platform, IoT layer, and the data layer. Ogiela and Ogiela (2014) proposed a cognitive system that could test a variety of diverse kinds of data and interpretations, which generate insights. CC can understand many tools, techniques, systems, and technologies, such as IoT, NLP, DL, ML, Big Data, and data processing.

Cognitive systems can learn, analyze, and remember a problem that might be contextually related to any type of organization. Learning capability of improvement without reprogramming and analyzing hypotheses based on current knowledge are the core characteristics of the cognitive system. A massive volume of data that is valuable for ML, such as DL and reinforcement learning methods in conjunction with CC, can provide solutions that are a significant symbiosis between CC and Big Data. Based on Park *et al.* (2019), these considerable aspects are scalability, natural interaction, and dynamism. In the Big Data analysis, the model adjustment is complete manually while in CC, the system allows incorporating changes alone. Based on Santos *et al.*'s (2017) research, when there are unstructured, structured, multimedia, and text data sources, the interpretation of the data set will enable a more precise meaning of a complex set of problems. These data and sources can be IoT tools, global positioning systems, social media, etc. (Chen *et al.*, 2017).

1.3.3 Cognitive Building

Cognitive buildings are becoming more common as technology develops. It gives project stakeholders tools they need to manage all stages of their projects better while additionally producing better experiences for their clients. Cognitive buildings' primary objective is to go further than the automation of the process and provide actionable insights with a more complex and integrated approach. Cognitive systems, by which advantages could be obtained from data analytic-based and self-learning abilities, have been experienced in various AEC industry fields, such as construction safety management, construction energy management, and intelligent business information management (Saurin *et al.*, 2008; Ogiela and Ogiela, 2014). Making the construction industry smarter and more intelligent saves time, cost, and energy, helps optimize the designs, and improves the safety, quality, and performance of projects. Consequently, buildings are shifting from predictive and adaptive to cognitive entities, interacting with the human activity within them through considerable advancements in the industry.

Among the whole advancements of technologies, the development of building automation allows all activities of projects which are aware of the conditions controlled within them and able to respond to various inputs. Once different required data is processed automatically and buildings respond accordingly, buildings are referred to as 'cognitive buildings' (Ploennigs *et al.*, 2017). Some of the significant CC features are reading, learning, reasoning, and making inferences from data that can be used in buildings (cognitive buildings) to ensure the comfortability and safety of occupants (Chen *et al.*, 2016). Cognitive buildings need to apply and utilize all accessible information and understand how it impacts building occupants. Some of the advantages of the interactive building include reduction of disruption; automated work order management/planning; predictive systems which decrease reactive maintenance; increasing use of space; enhancing user's comfort; easier problems' identification; streamlined maintenance management; integration with other software platforms to give building occupants an immersive technology experience; increase facility's value management by analytics processes; increase interaction with people to build the best space feasible, and enhance satisfaction. Gupta *et al.* (2018) presented the advantages of CC for a more expeditious Big Data analysis through utilization of the human brain, like computing power through AI. They showed that integrating CC and AI helps make quicker, better, and more accurate decisions and is more ideal for using Big Data.

1.4 Smart Cities and Digital Technologies

1.4.1 IoT-enabled Smart Cities

A smart city is an urban area that utilizes different varieties of IoT-based sensors to gather/produce data to use them efficiently and effectively to manage assets, resources, and services. Smart cities produce services that benefit from city-scale deployment of sensors, smart things, and actuators. The concept of a smart city combines ICT and various physical devices connected to the IoT network to optimize city operations and services' efficiency and connect to citizens. A big part of this ICT structure is an intelligent network of connected machines and objects transmitting data by wireless network technology and cloud-based systems. Cloud-based IoT applications obtain, analyze, and manage data in real-time to help municipalities, companies, and citizens and make more reliable decisions. Deakin and Al have listed four significant factors which contribute to the smart cities (Deakin and Al, 2011):

- The application of a wide range of electronic and digital technologies to cities.

- The use of ICT to transform life and working environments within the region.
- The embedding of such ICTs in government systems.
- The territorialization of practices brings people and ICTs unitedly to improve the knowledge and innovation they offer. The integration of a high number of connected devices and mobile interacting with technologies, sensors, users, and clouds requires a network that is more predictive, agile, and able to scale and move massive amounts of data in real-time rapidly and precisely.

Nowadays, numerous building systems, like the automation systems, the security systems, the energy management systems, and the computerized maintenance management system, are gathering a massive volume of data through advanced wireless sensors and emerging smart devices. Nevertheless, because of limitations in the inter-system data interoperability, the data formats vary based on various vendors. The evolving building systems already hold numerous valuable data for IoT-enabled smart city innovations. However, based on Gao *et al.* (2019), they need to set the integrated and federated building data foundation for optimum use of these systems. They also need to enable innovations in future smart cities by extracting relevant data from various building systems, storing them in cloud databases, and connecting these databases. Building Information Models (BIM) offer a powerful potential to improve the automation processes through IoT systems and CPS significantly. A strategy for linking BIM data schemas with the building automation and IoT data protocols can present a significant layer of spatial semantics to the IoT systems, like device geo-positioning and metadata tagging, and enrich efforts of smart buildings and cities while harmonizing these data sources with different data protocols.

1.4.2   CIoT-based Smart City Network Architecture

Min Chen *et al.* (2018) described the processes of interaction between machines and humans, enabling the machine to learn from cyberspace data and how diverse smart city-based applications and scientific experiments train through implementing AI and cognitive data. Their study provides a basic overview, benefits, and applications of applying human data to offer more personalized solutions by CC-based AI. In 2018, Alhussein et al. (2018) recommended a cognitive IoT/cloud-based smart healthcare framework that presented a smart city-based smart healthcare area application applying CC data. This study showed the possibility of gathering data from multiple sensors to train different smart city-based domain applications using CC. They also presented individual advantages, which can be achieved by using CC based on AI.

Smart city architecture in the context of IoT enhances citizens' security and quality of their life by providing more smart and accurate systems (Wei *et al.*, 2015; Fierro and Culler, 2015; Sharma *et al.*, 2018). Some of the newest advancements are aimed at facilitating the making of open platforms that simplify data integration and processing. Park *et al.* (2019) proposed a CIoT-based smart city network architecture, which contains many sub-layers, such as smart buildings, smart homes, smart energy, smart transportation, smart agriculture, and smart industry as seen in Fig. 1.3. In this platform, the Big Data generated from structured and unstructured data produces data gathered from sensors that contain multiple human aspects, like voice, emotions, brain activities. Smart buildings comprise numerous sensors that enable users to gather data from various sources to assistance-optimizing services in different sections, such as light, elevators, and HVAC quality control systems, optimizing energy consumption, etc. In the context of smart homes, IoT sensors acquire data regarding the users' movement, habitats, and their activities' schedule patterns, and their preferences set for indoor temperatures. The application of CC provides better control opportunities to enhance human safety and well-being.

Regarding smart energy, energy organizations provide advanced safety solutions to power plants by connecting the cognitive tool with imagery and weather analysis. In smart transportation, utilizing data obtained from human brain actions, environment, and emotions, a CC-based system is used through voice analysis to create immediate data on optimum directions and speed to serve for transportation to reach the destination on time. Automobiles and railway services have managed to implement solutions for more on-time and reliable services to their end-users. Different sensors in smart agriculture can gather rich-data-contained perceptual and logical activity and integrate them with imagery analysis to efficiently detect where the harvested fruits and vegetables are located and pick them. In the smart industry, integrating data gathered from brain sensors and environment sensors to cover human data industry performance can be optimized, more reliable/real-time decisions can be made, and raw data/resources can adequately be managed.

1.4.3　Smart Cities from a Machine Learning Perspective

Generating massive data on a city scale is a challenge for smart city designers and developers. As shown in Fig. 1.4, except Big Data management represented by volume, variety, and velocity (3Vs), there are other challenges from analytics and ML perspectives. So far, only a small fraction of the vast smart city data is typically used by smart services to increase the lives of the city's residents. Based on Mohammadi and Al-Fugaha (2018), the main culprit is the lack of a large amount of labeled data.

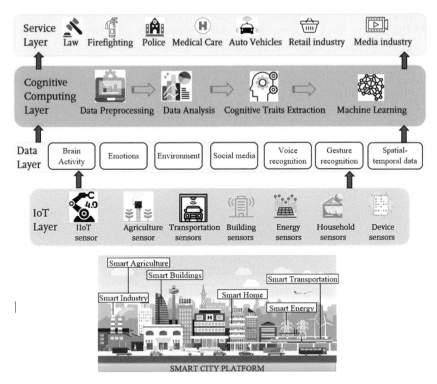

Fig. 1.3: CIoT-Net architecture (adapted from Park *et al.*, 2019)

This calls for the need to utilize ML algorithms that exploit the availability of unlabeled and labeled data in the context of smart cities. Smart city ecosystems from an ML standpoint have features, such as the need for interaction between humans and systems to provide their feedback; systems learn and improve themselves from previous experiences, and humans review this process; the need of systems for continuous learning mechanism as the context of a smart city application evolves; generated data by smart city applications has some degree of uncertainty.

Mohammadi and Al-Fuqaha (2018) introduced a framework for intelligence in smart cities, proposing three levels of intelligence consisting of the level of smart city/IoT infrastructure, cloud computing, and fog computing. In this framework, the overall position of ML methodologies for smart city infrastructure has each component of the smart city system which is controlled by an intelligent software agent that is deployed in the fog or the cloud depending on the characteristics of the required analytics. The main goal of this framework is that more profound data abstraction and knowledge representation can be taken as the data travels within the smart city infrastructure. At the highest level, a citywide consideration is required to manage the city's services and resources on a long-term basis. At the lowest level, smart objects and sensors that generate data are used to manage the services and resources on a short-term basis. Furthermore, fog-based analytics support local actions in predefined contexts, while cloud-based analytics can cover more significant geographical areas, including various contexts.

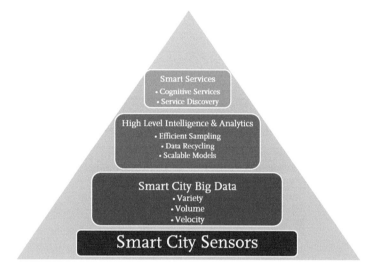

Fig. 1.4: Challenges of smart cities from a machine learning perspective (adapted from Mohammadi and Al-Fuqaha, 2018)

1.5 Conclusion

Potential applications of the CPSs, IoT, and CC for the AEC industry are numerous and diverse, providing various benefits for all activities done by stakeholders, firms, and the community as a whole. CPS can deliver construction stakeholders with the capability to gather data in an automated way, to simulate various scenarios, to perform progressive analysis, to visualize the simulations and evaluate results, and to control the equipment. A cognitive system, comprising ML, DL, and AI, utilizes Big Data created from the IoT network to develop a smart city platform in which cognitive buildings are connected to each other. This chapter demonstrates that CIoT-based smart city platform consisting of cognitive buildings equipped with smart facilities that generate a specific volume of data for the CIoT-based smart city network in real-time to contribute to the development of Construction 4 besides creation of a smart built environment. Future research challenges include human integration into CIoT-based smart city applications, i.e. from CPS to human CPS, to facilitate a completely autonomous society.

References

Abanda, F.H., Tah, J.H. and Keivani, R. (2013). Trends in built environment semantic Web applications: Where are we today? *Expert Systems with Applications*, 40(14): 5563-5577.

Alhussein, M., Muhammad, G., Hossain, M.S. and Amin, S.U. (2018). Cognitive IoT-cloud integration for smart healthcare: Case study for epileptic seizure detection and monitoring. *Mobile Networks and Applications*, 23(6): 1624-1635.

Alizadehsalehi, S., Hadavi, A. and Huang, J.C. (2020). From BIM to extended reality in AEC industry. *Automation in Construction*, 116: 103254.

Alpaydin, E. (2020). *Introduction to Machine Learning*, MIT Press.

Bedeley, R.T., Ghoshal, T., Iyer, L.S. and Bhadury, J. (2018). Business analytics and organizational value chains: A relational mapping. *Journal of Computer Information Systems*, 58(2): 151-161.

Borangiu, T., Trentesaux, D., Leitão, P., Boggino, A.G. and Botti, V. (2019). Service-oriented, holonic and multi-agent manufacturing systems for industry of the future. Proceedings of SOHOMA, Springer.

Breur, T. (2016). *Statistical Power Analysis and the Contemporary 'Crisis' in Social Sciences*, Springer.

Capuano, N. and Toti, D. (2019). Experimentation of a smart learning system for law based on knowledge discovery and cognitive computing. *Computers in Human Behavior*, 92: 459-467.

Chen, M., Herrera, F. and Hwang, K. (2018). Cognitive computing: Architecture, technologies and intelligent applications. *IEEE Access*, 6: 19774-19783.

Chen, M., Yang, J., Zhu, X., Wang, X., Liu, M. and Song, J. (2017). Smart home 2.0: Innovative smart home system powered by botanical IoT and emotion detection. *Mobile Networks and Applications*, 22(6): 1159-1169.

Chen, Y., Argentinis, J.E. and Weber, G. (2016). IBM Watson: How cognitive computing can be applied to Big Data challenges in life sciences research. *Clinical Therapeutics*, 38(4): 688-701.

Chiarello, F., Trivelli, L., Bonaccorsi, A. and Fantoni, G. (2018). Extracting and mapping Industry 4.0 technologies using Wikipedia. *Computers in Industry*, 100: 244-257.

Correa, F. and Maciel, A. (2018). A methodology for the development of interoperable BIM-based cyber-physical systems, ISARC. Proceedings of the International Symposium on Automation and Robotics in Construction, IAARC Publications.

Correa, F.R. (2018). Cyber-physical systems for construction industry. *IEEE Industrial Cyber-Physical Systems (ICPS)*, IEEE.

Erboz, G. (2017). How to define Industry 4.0: Main pillars of Industry 4.0. Szent Istvan University, *Gödöllő*: 1-9.

Faheem, M., Shah, S.B.H., Butt, R.A., Raza, B., Anwar, M., Ashraf, M.W., Ngadi, M.A. and Gungor, V.C. (2018). Smart grid communication and information technologies in the perspective of Industry 4.0: Opportunities and challenges. *Computer Science Review*, 30: 1-30.

Fierro, G. and Culler, D.E. (2015). Xbos: An extensible building operating system. Proceedings of the 2nd ACM International Conference on Embedded Systems for Energy-Efficient Built Environments.

Gao, X., Pishdad-Bozorgi, P., Shelden, D.R. and Tang, S. (2019). A scalable cyber-physical system data acquisition framework for the smart built environment. *Computing in Civil Engineering 2019: Smart Cities, Sustainability and Resilience*, American Society of Civil Engineers Reston, VA: 259-266.

Gao, X., Tang, S., Pishdad-Bozorgi, P. and Shelden, D. (2019). Foundational Research in Integrated Building Internet of Things (IoT) Data Standards.

Greer, C., Burns, M., Wollman, D. and Griffor, E. (2019). Cyber-physical systems and internet of things. *NIST Special Publication*, 1900: 202.

Gries, S., Meyer, O., Wessling, F., Hesenius, M. and Gruhn, V. (2018). Using blockchain technology to ensure trustful information flow monitoring in CPS. IEEE International Conference on Software Architecture Companion (ICSA-C), IEEE.

Group, I.W. (2013). Securing the future of German manufacturing industry— Recommendations for implementing the strategic initiative. Acatech, Munich, Germany, accessed Aug. 28, 2016.

Gupta, S., Kar, A.K., Baabdullah, A. and Al-Khowaiter, W.A. (2018). Big Data with cognitive computing: A review for the future. *International Journal of Information Management*, 42: 78-89.

ISO/IEC (2015). Smart cities. Preliminary Report, 2014, ISO/IEC.

Jacobsson, M., Linderoth, H.C. and Rowlinson, S. (2017). The role of industry: An analytical framework to understand ICT transformation within the AEC industry. *Construction Management and Economics*, 35(10): 611-626.

Jazdi, N. (2014). Cyber-physical systems in the context of Industry 4.0. 2014 IEEE International Conference on Automation, Quality and Testing, Robotics, IEEE.

Kambatla, K., Kollias, G., Kumar, V. and Grama, A. (2014). Trends in Big Data analytics. *Journal of Parallel and Distributed Computing*, 74(7): 2561-2573.

Klinc, R. and Turk, Ž. (2019). Construction 4.0 digital transformation of one of the oldest industries. *Economic & Business Review*, 21(3): 393-410. https://doi.org/10.15458/ebr.92

Lanko, A., Vatin, N. and Kaklauskas, A. (2018). Application of RFID combined with blockchain technology in logistics of construction materials. Matec Web of Conferences, EDP Sciences.

Lee, J., Bagheri, B. and Kao, H.-A. (2015). A cyber-physical systems architecture for Industry 4.0-based manufacturing systems. *Manufacturing Letters*, 3: 18-23.

Linares, D.A., Anumba, C. and Roofigari-Esfahan, N. (2019). Overview of supporting technologies for cyber-physical systems implementation in the AEC industry. pp. 495–504. *In*: Cho, Y.K., Leite, F., Behzadan, A., Wang, C. (Eds.). Computing in Civil Engineering 2019: Data, Sensing, and Analytics. American Society of Civil Engineers: Reston, VA, USA,

for Cyber-Physical Systems Implementation in the AEC Industry. In 2019; pp. 495–504.

Marques, M., Agostinho, C., Zacharewicz, G. and Jardim-Gonçalves, R. (2017). Decentralized decision support for intelligent manufacturing in Industry 4.0. *Journal of Ambient Intelligence and Smart Environments*, 9(3): 299-313.

Mohammadi, M. and Al-Fuqaha, A. (2018). Enabling cognitive smart cities using Big Data and machine learning: Approaches and challenges. *IEEE Communications Magazine*, 56(2): 94-101.

Muhuri, P.K., Shukla, A.K. and Abraham, A. (2019). Industry 4.0: A bibliometric analysis and detailed overview. *Engineering Applications of Artificial Intelligence*, 78: 218-235.

Nunes, D.S., Zhang, P. and Silva, J.S. (2015). A survey on human-in-the-loop applications towards an internet of all. *IEEE Communications Surveys & Tutorials*, 17(2): 944-965.

Ogiela, L. and Ogiela, M.R. (2014). Cognitive systems for intelligent business information management in cognitive economy. *International Journal of Information Management*, 34(6): 751-760.

Park, J.-H., Salim, M.M., Jo, J.H., Sicato, J.C.S., Rathore, S. and Park, J.H. (2019). CIoT-Net: A scalable cognitive IoT-based smart city network architecture. *Human-centric Computing and Information Sciences*, 9(1): 29.

Pishdad-Bozorgi, P., Gao, X. and Shelden, D.R. (2020). Introduction to cyber-physical systems in the built environment. pp. 22-40. *In*: Sawhney, A., Riley, M. and Irizarry, J. (eds). Construction 4.0: An Innovation Platform for the Built Environment. New York: Routledge.

Ploennigs, J., Ba, A. and Barry, M. (2017). Materializing the promises of cognitive IoT: How cognitive buildings are shaping the way. *IEEE Internet of Things Journal*, 5(4): 2367-2374.

Preissl, R., Wong, T.M., Datta, P., Flickner, M., Singh, R., Esser, S.K., Risk, W.P., Simon, H.D. and Modha, D.S. (2012). Compass: A scalable simulator for architecture for cognitive computing. SC'12: Proceedings of the International Conference on High Performance Computing, Networking, Storage and Analysis, IEEE.

Rezgui, Y., Boddy, S., Wetherill, M. and Cooper, G. (2011). Past, present and future of information and knowledge sharing in the construction industry: Towards semantic service-based e-construction. *Computer-aided Design*, 43(5): 502-515.

Santos, M.Y., E. Sá, J.O., Andrade, C., Lima, F.V., Costa, E., Costa, C., Martinho, B. and Galvão, J. (2017). A Big Data system supporting Bosch braga Industry 4.0 strategy. *International Journal of Information Management*, 37(6): 750-760.

Saurin, T.A., Formoso, C.T. and Cambraia, F.B. (2008). An analysis of construction safety best practices from a cognitive systems engineering perspective. *Safety Science*, 46(8): 1169-1183.

Sharma, P.K., Rathore, S. and Park, J.H. (2018). DistArch-SCNet: Blockchain-based distributed architecture with hi-fi communication for a scalable smart city network. *IEEE Consumer Electronics Magazine*, 7(4): 55-64.

Sheth, A. (2016). Internet of things to smart IoT through semantic, cognitive and perceptual computing. *IEEE Intelligent Systems*, 31(2): 108-112.

Smit, J., Kreutzer, S., Moeller, C. and Carlberg, M. (2016). Industry 4.0: Study, European Parliament.

Teizer, J., Allread, B.S., Fullerton, C.E. and Hinze, J. (2010). Autonomous pro-active real-time construction worker and equipment operator proximity safety alert system. *Automation in Construction*, 19(5): 630-640.

Uhlemann, T.H.-J., Lehmann, C. and Steinhilper, R. (2017). The digital twin: Realizing the cyber-physical production system for Industry 4.0. *Procedia Cirp*, 61: 335-340.

Vaidya, S., Ambad, P. and Bhosle, S. (2018). Industry 4.0: A glimpse. *Procedia Manufacturing*, 20: 233-238.

Wei, L., Yong-feng, C. and Ya, L. (2015). Information systems security assessment based on system dynamics. *Int. J. Secur. Appl.*, 9: 73-84.

Woodhead, R., Stephenson, P. and Morrey, D. (2018). Digital construction: From point solutions to IoT ecosystem. *Automation in Construction*, 93: 35-46.

Xu, J., Lu, W., Xue, F. and Chen, K. (2019). Cognitive facility management: Definition, system architecture and example scenario. *Automation in Construction*, 107: 102922.

Young, T., Hazarika, D., Poria, S. and Cambria, E. (2018). Recent trends in deep learning-based natural language processing. *IEEE Computational Intelligence Magazine*, 13(3): 55-75.

Zhang, M., Zhao, H., Zheng, R., Wu, Q. and Wei, W. (2012). Cognitive internet of things: Concepts and application example. *International Journal of Computer Science Issues (IJCSI)*, 9(6): 151.

Zhang, S., Yao, L., Sun, A. and Tay, Y. (2019). Deep learning based recommender system: A survey and new perspectives. *ACM Computing Surveys (CSUR)*, 52(1): 1-38.

Zheng, Y. (2018). Design and testing of automatic control system of intelligent building, 2018. 10th International Conference on Measuring Technology and Mechatronics Automation (ICMTMA), IEEE.

Towards a Digital Twin-based Smart Built Environment

Ibrahim Yitmen[1]* and Sepehr Alizadehsalehi[2]

[1] Department of Construction Engineering and Lighting Science, School of Engineering, Jönköping University, Gjuterigatan 5, 551 11, Jönköping, Sweden
[2] Project Management Program, Department of Civil and Environmental Engineering, Northwestern University, Evanston, IL, USA

1.1 Introduction

With the creation of massive data in projects (Big Data), the digitization/ computerization of various stages and processes of built environments are emerging to have a broad influence on how the architecture, engineering, and construction (AEC) projects are planned, built, and managed. The AEC industry moves to the digital era and smart industry to construct a smart built environment by utilizing Building Information Modeling (BIM), IoT, Information and Communications Technologies (ICT), cognitive computing (CC), cloud-based systems, Digital Twin (DT), and blockchain technologies. This chapter presents a comprehensive review and summary of the state-of-the-art research about Big Data, Big Data with CC, DT, DT in smart cities, BIM with DT, and blockchain with DT. This study provides a roadmap for future efforts toward implementing DT-based management systems in all phases of AEC industry applications. Therefore, the convergence of these emerging technologies with DT endorses the development of smart built environment.

2.2 Big Data

According to a published article by NASA scientists in 1997, large volumes of data have increased the demand for bigger memory space

*Corresponding author: ibrahim.yitmen@ju.se

in the local and remote disks and need for more resources. They called this issue a Big Data issue (Press *et al.*, 2016). In the last few years, more data has been created than in the entire history of human race. These data originated from various resources, such as images, voices, videos, social media information, financial information, location data, engineering information, etc. (Madanayake and Egbu, 2019). However, only a small portion of that generated data is ever used and analyzed. Big Data refers to a massive amount of structured, semi-structured, and unstructured data created by data-capturing technologies, which could not be acquired, stored, computed, analyzed, managed, and shared via conventional data systems within an adequate time and budget (Qi and Tao, 2018). These processes are beyond the conventional processing abilities of users. Big Data has attracted enormous attention of researchers and different industries as it has vast potential, and it can be applied in several high-impact areas, such as supply chain and logistics, operations management, decision making, e-commerce, sustainability, e-government, healthcare, market intelligence, and security (Kumar *et al.*, 2020).

2.1.1 Features of Big Data

Big Data has five key features – Variety, Volume, Velocity, Value, and Veracity (the 5Vs) (Wamba *et al.*, 2015; L'heureux *et al.*, 2017) as shown in Fig. 2.1.

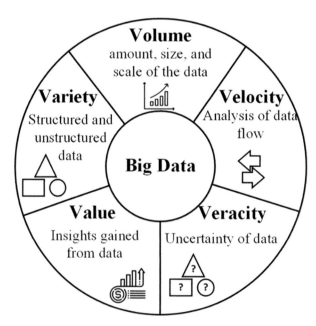

Fig. 2.1: The 5Vs of Big Data

- The term 'variety' in Big Data refers to all structured and unstructured data, which, generated by either humans or machines, and in data types, frequently require distinct processing capabilities and specialist algorithms (Hofmann, 2017). The most popular generated data under the variety category are structured, such as pictures, videos, texts, and unstructured data, such as voicemails, audio recordings, emails, hand-written texts, etc.
- The volume is one of the most critical characteristics of Big Data, which refers to the size, amount, and scale of the data that need to be analyzed and processed and typically are larger than terabytes and petabytes (Zhou *et al.*, 2016). The volume of data in Big Data is too large to process with traditional storage and processing capabilities and regular laptop or desktop processor and requires distinct and different processing technologies.
- The next dimension of Big Data is velocity, which refers to the speed with which data are being generated and analyzed (Yang *et al.*, 2017). Because of these data's high-velocity, they must be processed with advanced analytics and algorithms to reveal meaningful information (Hart, 2019).
- Veracity refers to the quality and accuracy of the analyzed data that needs processes to keep the low-veracity data from accumulating in the system. Low-veracity data includes a high rate of insignificant data (Dering and Tucker, 2017). The non-valuable in these data sets is referred to as noise. Veracity, in turn, represents the unreliability inherent in some data sources, which requires analysis to gain reliable predictions.
- Value refers to the extent to which Big Data generates worthwhile insights and benefits through data analysis (Holzwarth *et al.*, 2019). The essential element of the Big Data helps make sure that the data containing the required value is value (Brinch, 2018). The value is the end game of Big Data, which includes a large volume and variety of data that is easy to access and delivers quality analytics, enabling informed decisions.

2.1.2 Big Data Challenges for Cognitive Buildings

The cloud server utilizing IoT and the processing algorithms is projected to operate a significant amount of heterogeneous live data that are stored by numerous sensors at the cognitive buildings. The data also includes the operation and maintenance activities of workflow, the asset inventories, and dynamic environmental conditions at the buildings. In a cloud-based system design, the handling of data would not be a concern. Nevertheless, Big Data analysis in favor of decision-making needs is a challenging task to tackle in real-time. Big Data is capable of contributing to the cognitive buildings as it facilitates direct analysis of theoretical values, historical data, and the real condition of a building. Challenges that have arisen earlier can be learned, predictions can be achieved based on the plan, and proposals can be stipulated to circumvent similar problems in the future.

2.2 Digital Twins

2.2.1 Definitions of Digital Twins

With the evolution of Big Data and model-based engineering, the next logical step is to introduce the concept of DTs, extending model-based paradigms along the complete lifecycle. The theory of using Twins arises from National Aeronautics and Space Administration (NASA's) Apollo program, and the term DT was initially launched into NASA's integrated technology roadmap (Talkhestani *et al.*, 2019) and following appearing in product manufacturing (Boschert and Rosen, 2016; Schleich *et al.*, 2017) and more recently into smart cities and AEC industry (Mohammadi and Taylor, 2017; Howell and Rezgui, 2018). A DT refers to a digital replica of potential and actual physical assets (physical twin), devices, systems, processes, places, and people that can be used for different aims (El Saddik, 2018). Diverse definitions of DT have been shown in Table 2.1.

2.2.2 Characteristics of Digital Twins

DT is a concept that creates a physical asset model for various reasons, and its data can continuously adjust to changes in the environment or operation, using real time sensory data. DT can predict the future of the corresponding physical assets, monitor, and recognize any potential issues with its real physical counterpart to allow the forecast of the remaining useful life of the physical twin by leveraging a combination of physics-based models and data-driven analytics (Tuegel *et al.*, 2011). The fundamental technology to move a DT that assists the flow of raw sensory data to high-level understanding and insights information is the data and information fusion (Liu et al. 2018). Figure 2.2 shows the characteristics and features of DT in general.

The essential functionality of DT implementation is to provide accurate operational pictures of the assets, allowing the DT to mirror the activities of its corresponding physical twin with the capabilities of prediction, anomaly detection, early warning, and optimization. Based on Kaur *et al.* (2020), DT consists of three (3) key components of physical products in real space, virtual products in virtual space, and the connections of data and information that will tie the virtual and real products together. The underlying architecture of DT consists of sensor and measurement technologies, the Internet of Things (IoT), and cognitive computing. As shown in Fig. 2.3, the IoT technologies and related systems act to collect real-time throughout the edge computing devices and the smart gateway,

Table 2.1: A Selection of Definitions of DTs Based on Academic Publications

No	References	Year	Application area
1	Bolton, A., 2018	2018	"A realistic digital representation of assets, processes or systems in the built or natural environment."
2	Tao, Sui *et al.*, 2019	2018	"DT is a real mapping of all components in the product lifecycle using physical data, virtual data, and interaction data between them."
3	Bolton, McColl-Kennedy *et al.*, 2018	2018	"A dynamic virtual representation of a physical object or system across its lifecycle, using real-time data to enable understanding, learning, and reasoning."
4	El Saddik, 2018	2018	"A DT is a digital replica of a living or non-living physical entity. By bridging the physical and the virtual world, data is transmitted seamlessly, allowing the virtual entity to exist simultaneously with the physical entity."
6	Söderberg, Wärmefjord *et al.*, 2017	2017	"Using a digital copy of the physical system to perform real-time optimization."
7	Grieves and Vickers, 2017	2017	"The Digital Twin is a set of virtual information constructs that fully describe a potential or actual physical manufactured product from the micro atomic level to the macro geometrical level. At its optimum, any information that could be obtained from inspecting a physically manufactured product can be obtained from its DT."

allowing things and people to be connected to anything, anyone, at anytime, and anywhere using any service, network, or path (Atlam and Wills, 2020).

IoT enables access and connection to intelligence data and is interlinked with DTs and digital models, which virtually represent their physical counterparts (Tang, Shelden *et al.*, 2019). As an umbrella term, cognitive computing consists of technologies, including Big Data, machine learning, cognitive algorithms. AI tries to simulate human thought processes in the computer model through DT technology (Xu, Lu *et al.*, 2019). Cognitive computing is replicating the way users are analyzing and managing data in the physical world (Alizadehsalehi *et al.*, 2020). DT uses particular

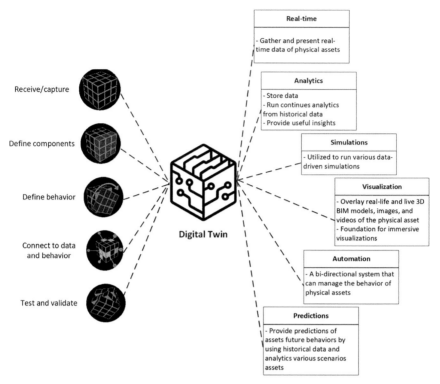

Fig. 2.2: Characteristics and features of DT in general

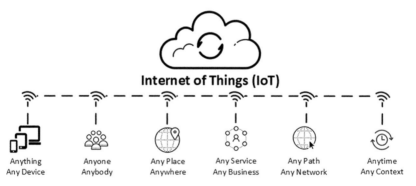

Fig. 2.3: The IoT can connect anything anywhere using any path
(adapted from Atlam and Wills, 2020)

simulation models based on their ability to describe behavior, solve issues, and derive relevant solutions for real-life systems (Khajavi *et al.*, 2019).

DTs serve to define the optimal set of parameters and actions which support and maximize any of the main performance metrics (KPIs) and present predictions for long-term outlining. It can be used in the healthcare industry to capture and visualize a hospital system to produce safe conditions and test the influence of possible system performance changes. Beyond the operations also, it helps to increase the quality of health services performed to the patients (Bagaria *et al.*, 2020). In the automotive industry, the DT of the product contains the entire car, its physical behavior, software, electrics, and mechanics (Augustine, 2020). DT allows to simulate and validate each step of the development to identify issues and potential failures beforehand.

2.2.3 Applications of Digital Twins

The DT has a high potential to be used in diverse fields and industries (Daily and Peterson, 2017). In 2018, DT was between the top ten strategic technology trends. Based on future research forecasts, the DT market will touch fifteen (15) billion dollars by 2023 (El Saddik, 2018). With the increasing deployment of IoT-based systems and technologies and cognitive computing sciences, the idea of a digital version of all physical things has gained momentum, and it is ideal for businesses to leverage DT platforms to advance their services and principles. Such a DT-driven product design, manufacturing, and service with Big Data have numerous applications in various industries, such as aerospace, healthcare, smart cities, AEC industry, manufacturing, agriculture, etc. (Fig. 2.4).

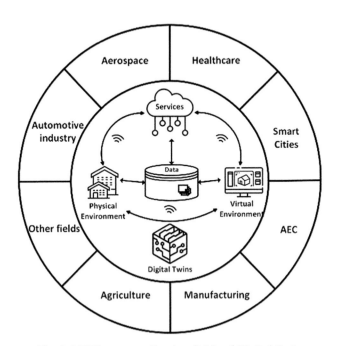

Fig. 2.4: Different application fields of Digital Twin

DTs are widely acknowledged in both industry and academic publications, mostly in the manufacturing industry, smart cities, and currently in the AEC industry. Table 2.2 summarizes previous research that used DT technologies in different domains of the AEC industry, manufacturing, and smart city from 2017 through 2020.

DT in the manufacturing industry is very popular and has numerous applications. The main reason is that manufacturers are always looking for efficient and effective methods and technologies to increase the quality of their products, decrease production time, reduce cost, and motivate their stakeholders (Mi *et al.*, 2019). This growth and maturity align with the Industry 4.0 concept, which coined the fourth industrial revolution. This provides the connectivity of devices to make DT a reality for manufacturing processes (Mawson and Hughes, 2019). The DT provides the manufacturer with the ability to check real-time status on machine performance, production line feedback, and predict issues sooner. Use of DT in manufacturing enhances connectivity and feedback between devices, improving reliability, and performance. DT, with the help of AI and cognitive computing sciences have multiple potentials and advantages for the manufacturing industry, such as:

- higher precision as the machines can hold higher volumes of data, required for prediction and performance analysis
- building an environment to test products and a system which acts on real-time data, within a manufacturing setting increases the automation of the industry improves data analytics and simulation
- improves the accuracy and efficiency of testing as it can perform data analytics on live vehicle data to forecast the performance of current and future components

2.2.4 Role of Digital Twins in Smart Cities

The smart city area is another sector for the application of DT technology and its potential in this industry shows fast improvements in connectivity through IoT. The number of smart cities is rapidly increasing around the world, and they require more connected devices and technologies to capture, transfer, and save various types and volumes of data and information (Lu *et al.*, 2020). Moreover, DT in smart cities can analyze performance data gathered over time and under varied circumstances, capturing the special and temporal implications of optimizing urban sustainability. Furthermore, DT in smart city has other benefits, such as:

- planning and development of current smart cities
- energy-saving
- utility management and how they are distributed and used
- create a living testbed within a virtual twin by testing scenarios and

Table 2.2: Digital Twin Technology in Various Domains of Industry, 2017-2020

No	Reference/Year	Domain	Evaluation methodology	Focus area
1	Boje, Guerriero et al., 2020	AEC industry	Review	BIM, IoT, AI, Big Data
2	Lu, Xie et al., 2020	AEC industry	Case study	IFC, Operation and Maintenance
3	Francisco, Mohammadi et al., 2020	Smart City	Case study	Energy management, Commercial buildings
4	Mateev, 2020	AEC industry	Concept	IoT, Industry 4.0, Microsoft Azure
5	Lu, Parlikad et al., 2020	Smart City	Case study	Asset management, Operation and maintenance, Building and city levels
6	Bilberg and Malik, 2019	Manufacturing	Case study	Simulation
7	Howard, 2019	Manufacturing	Concept	EDA, Visualization
8	Karadeniz, Arif et al., 2019	Manufacturing	Case study	AR, VR, AI, CPS, Industry 4.0
9	Lu, Xie et al., 2019	AEC industry	Concept	BIM, Asset management, Operation and Maintenance
10	Lu, Peng et al., 2019	Manufacturing	Review	CPS, Cloud, Industry 4.0
11	Mandolla, Petruzzelli et al., 2019	Manufacturing	Case study	Blockchain, Visualization
12	Mawson and Hughes, 2019	Manufacturing	Case study	Industry 4.0
13	Khajavi, Motlagh et al., 2019	AEC industry	Case study	BIM, IoT, Lifecycle management, Wireless sensor network
14	Min, Lu et al., 2019	Manufacturing	Case study	AI, Optimization

(Contd.)

Table 2.2: (*Contd.*)

No	Reference/Year	Domain	Evaluation methodology	Focus area
15	Shangguan, Chen *et al.*, 2019	Manufacturing	Case study	CPS
16	Tao, Qi *et al.*, 2019	Manufacturing	Review	AI, CPS, Industry 4.0
17	Xu, Sun *et al.*, 2019	Manufacturing	Concept	CPS, AI, Industry 4.0, Transfer Learning
18	Ruohomäki, Airaksinen *et al.*, 2018	Smart City	Case study	Sensor Ontology, Visualization
19	Kaewunruen and Xu, 2018	AEC industry	Case study	BIM, Sustainability, Railway station buildings
20	Mohammadi and Taylor, 2017	Smart City	Concept	Simulation, VR

learning from the environment by analyzing changes in the data collected
• increase in data analytics and monitoring quality

Figure 2.5 shows a sample of DT, the creation system, and different environments like hospitals, schools, offices, banks, warehouses, stadiums, factories, streets, parking lots, etc.

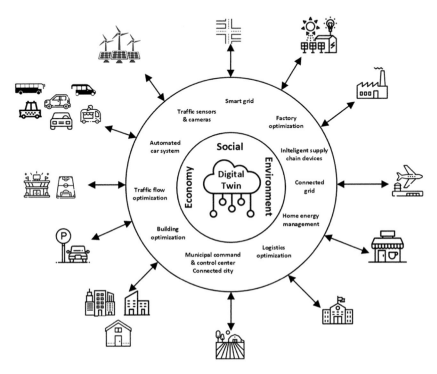

Fig. 2.5: Azure Digital Twins (adapted from Azure, 2020)

2.3 Digital Twins in AEC Industry

2.3.1 Evolution from BIM to Digital Twins

The AEC industry is another sector that hosts a variety of applications for DT usage. In the last few years, BIM paradigm has improved collaboration during all stages of construction projects in design and construction, and it has quickly shifted to one of the most popular topics in adjacent research fields over the built environment lifecycle, at any sizes and types of building, infrastructure, and city levels (Alizadehsalehi and Yitmen, 2018; Salehi and Yitmen, 2018). BIM was initially envisaged to expedite the effective and efficient generation, exchange, and manage accurate data/information (Alizadehsalehi and Yitmen, 2016). However, it now faces significant challenges where big leveraging data are heralded as potential solutions to automation (Alonso *et al.*, 2019). The progression of BIM interoperability solutions, from International Organization for Standardization Standard for the Exchange of Product model data (ISO-STEP) to Industry Foundation Classes (IFC) and more recently, IfcOwl are seemingly unable to efficiently leap from a static BIM to a web-based paradigm (Pauwels *et al.*, 2017). For an efficient and effective exchange of data and information among BIM software and platforms, the AEC industry currently faces significant challenges wherever leveraging Big Data, IoT, and AI are heralded as potential solutions to automation and brother applications. Furthermore, generated data in the design and construction phases of projects has increased since BIM adoption, experiencing what is termed 'drowning in data', using little added benefit to the construction supply chain to date (Boje *et al.*, 2020).

Appearing BIM tools and technologies have slowly shifted the process data and information on the AEC projects and it is generated, stored, managed, and exchanged between all involved stakeholders in different phases of projects. The conception of IFC alone has had a vital influence on how current tools and techniques are produced in AEC research and development (R&D) (Howell and Rezgui, 2018). However, digital-based technologies across the industry, taking advantage of the IoT and AI sciences, such as data analytics, ML, DL, etc., are advancing at an ever-increasing pace. Therefore, the evolution of BIM should be precisely drafted within a paradigm that includes people, processes, and emerging technologies in a more interconnected industry (Batty, 2018).

2.3.2 Concept of Digital Twins

Digitalization and computerization of data and information flow have a broad influence on the system and process of managing the lifecycle of projects. These data must be captured accurately, stored safely, and shared securely, and technologies require to be well designed to ensure efficiency and security (Lu *et al.*, 2019). Consequently, the concept of DT was introduced to comprehensively store, manage, analyze, predict, plan, and present all project data. The DT is widely promoted in AEC industry. DT is a digital model that is a dynamic representation of an asset and mimics its real-world behaviors (Commission, 2017). DT is built on data. DTs in AEC projects can significantly increase all phases and stages of any type and size of project during its lifecycle, such as design, simulation, planning, construction, monitoring, operating, maintaining, optimizing, and demolition (Oliver *et al.*, 2018; Mateev, 2020). One of DT's main steps is to transfer the collected data to the higher layers for modeling and analyzing. Multiple communication tools and technologies could be applied in this layer, like short-range coverage access network technologies, such as WiFi, Zigbee, near-field communication (NFC), mobile-to-mobile (M2M), and Zwave; and broader coverage, such as 3G, 4G, long-term evolution (LTE), 5G, and low-power wide-area networks (LP-WAN) (Lu *et al.*, 2020).

The concept of DT is gaining currency in the AEC industry. DT is an up-to-date and dynamic model of a physical asset or facility. It includes all the structured and unstructured information about projects that can be shared among team members. DTs help the AEC industry to model, simulate, understand, predict, and optimize all aspects of a physical asset or facility. For the design phase, it is possible to virtually create a solution and accurately render it operational before a single physical action is taken. In design, a DT is used to create the optimum solution. It results from exhaustive simulations and rich data that specify areas, such as best architecture, configuration, materials, and cost. Then it can simulate that solution under different types of real scenarios. Based on the data provided, the design can then be modified. In the build phase, a DT can be used to provide the construction specifications or what are called parametric estimates to different providers. In this way, a DT can be an asset in streamlining the procurement process. Besides, and importantly, during the build, sensors are applied to the physical object to collect and transmit data back to its virtual replica. This enables to perform magic during the operational and maintenance phases. At this point, with enough sensors, the virtual twin is providing all relevant data about the state of the physical twin. For example, an MEP section or structural element of a building can render accurately in its DT its temperature, vibration, strength, and so much more. All of this becomes possible because of increasingly better digital technologies that include faster computers, better telemetry, that is, the communication of measurements from a collection point to receiving equipment, smaller, more accurate sensors, data management, and artificial intelligence.

Nasaruddin *et al.* (2018) demonstrated the DT of a building as the "interaction between the real-world building's indoor environment and a digital yet realistic virtual representation model of the building environment, which provides the opportunity to real-time monitoring and data acquisition". Furthermore, El Saddik (2018), in another research, showed an expanded explanation for DT: "DTs will facilitate the means to monitor, understand, and optimize the functions of all physical entities, living as well as non-living, by enabling the seamless transmission of data between the physical and virtual world." Some of the benefits of creating a DT for an indoor area of building projects are gathering, generating, and visualizing the building's environment, such as airflow, relative humidity, air temperature, and lighting condition and then analyzing data irregularities and optimizing building services (Khajavi *et al.*, 2019).

2.3.3 Capabilities of Digital Twins

During operations, an abundance of data is being collected and fed back to its DT over a digital thread. Backed by artificial intelligence, the DT can identify and even predict maintenance issues before they happen (Fig. 2.6). It has become a data-informed model of a physical system. This compelling feature reduces cost since it is typically cheaper to conduct maintenance than repair it after it faces issues proactively. Finally, this continuous real-time feed of data can help with optimization, that is, improve its performance by enabling the system to either automatically modify its behavior or prompt the manual intervention of a human. DTs have become particularly ubiquitous in the IoT world. These internet-connected electronics collect and produce data and services and interact and communicate with each other and the central systems. The data collected from these devices create detailed knowledge, enabling capabilities.

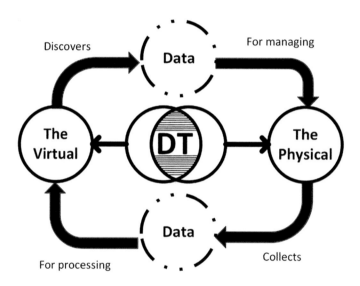

Fig. 2.6: The Digital Twin paradigm (adapted from Boje *et al.*, 2020)

With billions of new IoT devices being deployed and managed each year, it must be clear by now that DTs have a remarkable future ahead. In the world of DTs, data is king. By understanding the real-time data produced by the physical object, we can analyze its current state, its behavior over time, and even predict future scenarios. This can enable more efficient monitoring that helps, for example, in identifying potential problems. The real-time data also supports the implementation of system improvements and opens up options for adding new value. The accuracy of measured data entirely depends on how precise the captured data by various sensors are and capturing devices. Determining the number of variables to measure is based entirely on each use. As a minimum, a DT must provide the following five capabilities:

- *Connect*: There must be a 'live' connection between the digital replica and the physical world. This connection allows various disparate information and data from the physical world to come into a unified virtual environment.
- *Integrate*: Intelligently checks and links relevant data from different sources (and across sectors) to effectively enable a meaningful analysis to those who see the value.
- *Visualize*: To display real-time multisource data to the user. This allows access to the information users' need, precisely when they need it, across the whole project and asset operation lifecycle.
- *Analyze*: Federated data sets from various sources can be processed, modeled, analyzed, and simulated to bring business objectives to life.
- *Secure*: Having a security-minded management approach to data and information by applying relevant technical security and privacy standards.

Based on a report of Oracle in 2018 (Oracle, 2018), DT has seven values and applications in the AEC industry and these are as follows: (1) Real-time remote monitoring and control; (2) greater efficiency and safety when data required; (3) predictive maintenance and scheduling by intelligent data analysis; (4) a better risk assessment by a study of unexpected scenarios, the response of the operation, and the corresponding strategies of mitigations; (5) better collaboration through greater autonomy; (6) efficient, fast, and informed decision support by accessibility to quantitative data and advanced analytics in real-time for more informed and quicker decision-making; and (7) better communication and documentation with automated reporting through available information in real-time.

DT is a virtual representation of the elements and dynamics of IoT devices. It affects the design of the build and operations of how the product is pulled together. Analytics has to be real-time, operational, quality, and predictive oriented in its nature. The goal is to make a dynamic model that users can shift as they go through the design, build, and operation phases in the lifecycle of the product. The DT not only captures the engineering aspects, but it also captures the industry context, and the dynamics of how that product is used at the same time. DT can use the same product differently in different industries and has different results for that product based on how the industry uses it. For instance, a pump can be used in the oil and gas industry or the same pump can be used differently in the construction industry. The outcome is based on the industry context of how that device is going to be used.

2.3.4 Developing a City-level DT

The relation among the DT and smart cities has been recognized as the novel and ultimate technological apparatus for smartening cities. The potential for DTs to be dramatically capable within a smart city is increasing due to the fast-technological improvement of IoT and CC. The concept of large-scale DTs for the built environment has recently found widespread favor, which is enabled by the fast development of digital infrastructure to digitalize the buildings and cities. At the city level, due to the massive volume of data, the data collection process for DT is a challenging task, particularly while considering the format, type, source, and content of data.

Moreover, the sub-assets, like buildings will have their sub-DTs regarding their daily function services, and these sub-DTs will further provide essential data while receiving a query from the city DT. Some of the primary data acquisition technologies/techniques are contactless data capturing (image-based techniques and radio-frequency identification (RFID)), wireless communication, distributed sensor systems, and mobile access. Each twin is designed, based on the DT architecture and different levels of sub-DTs (i.e. buildings), including real-time data acquisition, efficient data integration, and management.

Figure 2.7 shows the development DT systems architecture for building and city levels. By sharing and using the same concept for city DTs, building DTs also include data acquisition layers (such as utilizing various types of wireless sensors, IoT-based devices, and quick response (QR) codes), transmission layer, digital modeling layer, data/model integration layer (such as data analysis functions and simulation engine), and service layer (such as workplace design and space utilization).

2.4 Digital Twins and Blockchain

2.4.1 Integration of Blockchain and Digital Twins

A blockchain is a growing list of blocks linked by using cryptography, in which any block comprises a cryptographic hash of the preceding block, transaction data, and a timestamp. It is a decentralized, distributed, and public digital ledger in which each involved record cannot be altered retroactively, without altering all succeeding blocks (Huang *et al.*, 2020). Blockchain has naturally become the next level in linking the connection. This technology redefines DTs' concept to aid the application of DTs in IoT, transferring the data and value onto the Internet with full transparency.

The protection and security of the DT platform and data are essential. DT and blockchain can be leveraged for their security features and help businesses thwart instances of fraud and duplication of their products

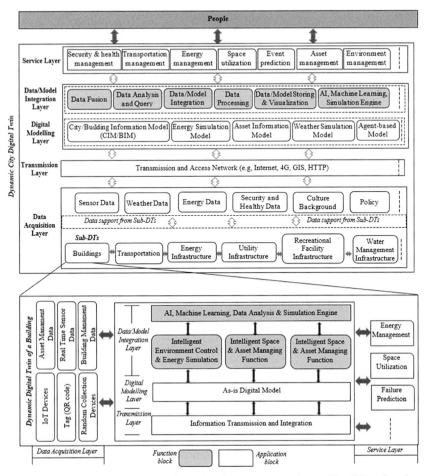

Fig. 2.7: System architecture of DT development at a city and building level (adapted from Lu *et al.*, 2020)

and services. Blockchain empowers the security of the DT system through cryptographic hashing algorithms. This process is possible by storing DT data on distributed ledgers that cannot be changed easily and cannot be controlled by any central authority (Yaqoob *et al.*, 2020). Immutability, as one of the main distinctive characteristics of blockchain, helps to ensure its ledger to remain indelible and keep the unalterable history of DTs transactions. These characteristics bring a unique level of trust, provide integrity of DTs' data, and change the DTs' audit process into a cost-effective and efficient procedure. Communication activities are needed by the participators from the same and various participant types. In each participant type, some participants are resulting in a complicated communication network. As shown in Fig. 2.8, creating a peer-to-

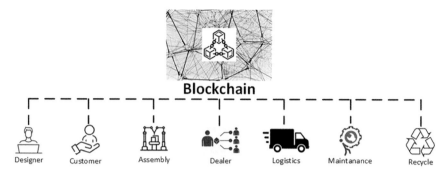

Fig. 2.8: Peer-to-peer network for data management (adapted from Huang *et al.*, 2020)

peer network needed through blockchain technology can enhance the communication effectively among the participants in the product lifecycle.

With the help of blockchain and without any third party from multiple locations in a secure manner, DTs can get access to the distributed databases. As the DTs contain blockchain, the DTs' information quickly and simply can be tracked/accessed. Therefore, provenance tracking, record keeping, and auditability becomes easy. Also, as the DTs' history is traceable easily by blockchain, this accurate provenance tracking can be applied to analyze and detect fraud in all parts of DTs. Blockchain enforces transparency and accountability through encryption and control mechanisms (e.g., smart contracts and chain code).

2.4.2 Creation of Digital Twin Process Using Blockchain

The DT creation process needs to be trusted in tracking and management to ensure which complete history of data and information is kept securely. Therefore, blockchain is used to meet strict security conditions in a decentralized method. DT-involved related stakeholders, such as process managers, phase managers, and owners interact with the smart contract through a front-end layer using Application Program Interfaces (APIs). The front-end decentralized application may use any of the interfaces, such as RestHTTP, Web3, or JSON RPC, to connect the stakeholders to the smart contract or Interplanetary File System (IPFS) servers. The smart contract interactions ensure secure on-chain resources that are traceable and tamper-proof (Hasan *et al.*, 2020). Figure 2.9 presents the main phases included in the creation process of DTs. There are mainly four phases of design, build, test, and delivery in creating a DT (Hasan, Salah *et al.*, 2020).

Li *et al.* (2019) examined the synergies among distributed ledger technology (DLT), such as blockchain, IoT, BIM, and smart contracts. They introduced an approach that connects the digital environment, the physical environment, the contract, and the DLT environment. To proof

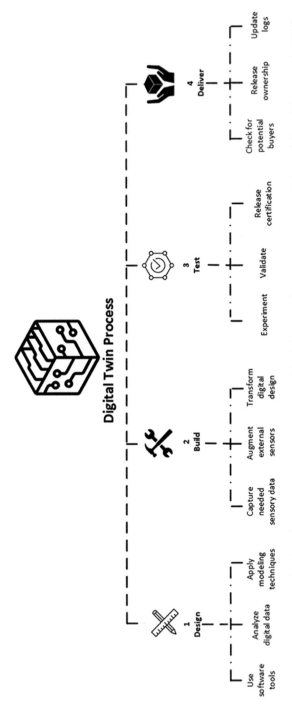

Fig. 2.9: Phases involved in the DT creation process using blockchain as the managing entity (adapted from Hasan *et al.*, 2020)

of their concepts, they used a simulated installation activity to verify the conceptual interrelations included in the recommended framework. Their research shows how a mini smart contract can perform within their created approach and automate the payment process. In the end, they discuss the main challenges and limitations of this approach.

2.5 Conclusion

BIM encounters key challenges where managing Big Data, IoT, and AI are designated as potential solutions to automation and the inclusion of broader environmental settings for the development of smart built environment. The necessity to monitor and control assets (manufactured elements, buildings, bridges, etc.) throughout their lifecycle, linked with advances in technological facilities, have shifted various research topics into exploring DT uses and potential. DT encompasses sensors, measurement technologies, IoT, simulation, and modeling and ML. IoT-cloud communication models, and Big Data created by IoT devices develop enhanced potential, and incremental data of cloud services to create a dynamic digital replica of buildings and city that integrates each sub-DT (e.g. building DT and bridge DT) in smart city ecosystem. Creating DTs on blockchain and compelling transparency and accountability through encryption and controlling mechanisms (e.g. smart contracts and chaincode) can be utilized to preserve IoT systems and devices, such as temperature sensors, security cameras, air quality sensors, etc. from probable attacks, as it facilitates devices to make security decisions without depending on a central authority in smart city ecosystem. The future research challenges include scalability, standardization and regulations, data privacy, and interoperability for the successful adoption of blockchain in DTs, alignment of DTs infrastructure with the smart IoT-enabled devices in such a manner that collaboration can be improved.

References

Alizadehsalehi, S., Hadavi, A. and Huang, J.C. (2020). From BIM to extended reality in AEC industry. *Automation in Construction*, 116: 103254.

Alizadehsalehi, S. and Yitmen. I. (2016). The Impact of Field Data Capturing Technologies on Automated Construction Project Progress Monitoring, Procedia Engineering, Elsevier.

Alizadehsalehi, S. and Yitmen, I. (2018). A concept for automated construction progress monitoring: Technologies adoption for benchmarking project performance control. *Arabian Journal for Science and Engineering*, 1-16.

Alonso, R., Borras, M., Koppelaar, R.H., Lodigiani, A., Loscos, E. and Yöntem, E. (2019). *Sphere: BIM Digital Twin Platform*. Multidisciplinary Digital Publishing Institute Proceedings.

Atlam, H.F. and Wills, G.B. (2020). IoT security, privacy, safety and ethics. *Digital Twin Technologies and Smart Cities*, Springer, 123-149.

Augustine, P. (2020). The industry use cases for the Digital Twin idea. *Advances in Computers*, Elsevier, 117: 79-105.

Azure (2020). https://azure.microsoft.com/en-us/blog/announcing-azure-digital-twins-create-digital-replicas-of-spaces-and-infrastructure-using-cloud-ai-and-iot/.

Bagaria, N., Laamarti, F., Badawi, H.F., Albraikan, A., Velazquez, R.A.M. and El Saddik, A. (2020). Health 4.0: Digital twins for health and well-being, *Connected Health in Smart Cities*, Springer, 143-152.

Batty, M. (2018). *Digital Twins*. SAGE Publications, UK: London, England.

Bilberg, A. and Malik, A.A. (2019). Digital Twin-driven human-robot collaborative assembly. *CIRP Annals*, 68(1): 499-502.

Boje, C., Guerriero, A., Kubicki, S. and Rezgui, Y. (2020). Towards a semantic construction Digital Twin: Directions for future research. *Automation in Construction*, 114: 103179.

Bolton, A., Enzer, M. and Schooling, J. (2018). The Gemini Principles: Guiding values for the national digital twin and information management framework. Centre for Digital Built Britain and Digital Framework Task Group.

Bolton, R.N., McColl-Kennedy, J.R., Cheung, L., Gallan, A., Orsingher, C., Witell, L. and Zaki, M. (2018). Customer experience challenges: Bringing together digital, physical and social realms. *Journal of Service Management*, 29(5): 776-808.

Boschert, S. and Rosen, R. (2016). Digital Twin – The simulation aspect. *Mechatronic Futures*, Springer, 59-74.

Brinch, M. (2018). Understanding the value of Big Data in supply chain management and its business processes. *International Journal of Operations & Production Management*, 38(7): 1589-1614.

Commission, N.I. (2017). Data for the public good. NIC Report.

Daily, J. and Peterson, J. (2017). Predictive maintenance: How Big Data analysis can improve maintenance. *Supply Chain Integration Challenges in Commercial Aerospace*, Springer, 267-278.

Dering, M.L. and Tucker, C.S. (2017). Generative adversarial networks for increasing the veracity of Big Data. 2017 IEEE International Conference on Big Data, IEEE.

El Saddik, A. (2018). Digital Twins: The convergence of multimedia technologies. *IEEE MultiMedia*, 25(2): 87-92.

Francisco, A., Mohammadi, N. and Taylor, J.E. (2020). Smart City Digital Twin-enabled energy management: Toward real-time urban building energy benchmarking. *Journal of Management in Engineering*, 36(2): 04019045.

Grieves, M. and Vickers, J. (2017). Digital Twin: Mitigating unpredictable, undesirable emergent behavior in complex systems. *Transdisciplinary Perspectives on Complex Systems*, Springer, 85-113.

Hart, J. (2019). *Modern Predictors of Gentrification: Utilizing High-velocity, Big Data*. http://viz.jackhart.georgetown.domains/Viz_class.pdf

Hasan, H.R., Salah, K., Jayaraman, R., Omar, M., Yaqoob, I., Pesic, S., Taylor, T. and Boscovic, D. (2020). A blockchain-based approach for the creation of Digital Twins. *IEEE Access*, 8: 34113-34126.

Hofmann, E. (2017). Big Data and supply chain decisions: The impact of volume, variety and velocity properties on the bullwhip effect. *International Journal of Production Research*, 55(17): 5108-5126.

Holzwarth, V., Schneider, J., Kunz, A. and vom Brocke, J. (2019). Data driven value creation in AEC along the building lifecycle. *Journal of Physics: Conference Series*, IOP Publishing.

Howard, D. (2019). The Digital Twin: Virtual Validation. *In:* Electronics Development and Design. 2019 Pan Pacific Microelectronics Symposium (Pan Pacific), IEEE.

Howell, S. and Rezgui, Y. (2018). *Beyond BIM: Knowledge Management for a Smarter Built Environment*. IHS Markit Publications: London, England.

Howell, S. and Rezgui, Y. (2018). *Beyond BIM: Knowledge Management for a Smarter Future*. IHS Markit.

Huang, S., Wang, G., Yan, Y. and Fang, X. (2020). Blockchain-based data management for Digital Twin of product. *Journal of Manufacturing Systems*, 54: 361-371.

Kaewunruen, S. and Xu, N. (2018). Digital Twin for sustainability evaluation of railway station buildings. *Frontiers in Built Environment*, 4: 77.

Karadeniz, A.M., Arif, İ., Kanak, A. and Ergün, S. (2019). Digital Twin of eGastronomic things: A case study for ice cream machines. 2019 IEEE International Symposium on Circuits and Systems (ISCAS), IEEE.

Kaur, M.J., Mishra, V.P. and Maheshwari, P. (2020). The convergence of Digital Twin, IoT, and machine learning: Transforming data into action. *Digital Twin Technologies and Smart Cities*, Springer, 3-17.

Khajavi, S.H., Motlagh, N.H., Jaribion, A., Werner, L.C. and Holmström, J. (2019). Digital Twin: Vision, benefits, boundaries, and creation for buildings. *IEEE Access*, 7: 147406-147419.

Kumar, N., Gayathri, N., Rahman, M.A. and Balamurugan, B. (2020). *Blockchain, Big Data and Machine Learning: Trends and Applications*. Boca Raton: CRC Press.

L'heureux, A., Grolinger, K., Elyamany, H.F. and Capretz, M.A. (2017). Machine learning with Big Data: Challenges and approaches. *IEEE Access*, 5: 7776-7797.

Li, J., Kassem, M., Ciribini, A.L.C. and Bolpagni, M. (2019). A proposed approach integrating DLT, BIM, IOT and smart contracts: Demonstration using a simulated installation task. pp. 275-282). *In*: International Conference on Smart Infrastructure and Construction 2019 (ICSIC) Driving Data-informed Decision-making. ICE Publishing.

Liu, Z., Meyendorf, N. and Mrad, N. (2018). The role of data fusion in predictive maintenance using digital twin. AIP Conference Proceedings, AIP Publishing, LLC.

Lu, Q., Xie, X., Heaton, J., Parlikad, A.K. and Schooling, J. (2019). From BIM Towards Digital Twin: Strategy and Future Development for Smart Asset Management. International Workshop on Service Orientation in Holonic and Multi-Agent Manufacturing, Springer.

Lu, Q., Parlikad, A.K., Woodall, P., Don Ranasinghe, G., Xie, X., Liang, Z., Konstantinou, E., Heaton, J. and Schooling, J. (2020). Developing a Digital

Twin at building and city levels: Case study of West Cambridge Campus. *Journal of Management in Engineering*, 36(3): 05020004.

Lu, Q., Xie, X., Parlikad, A.K. and Schooling, J.M. (2020). Digital Twin-enabled anomaly detection for built asset monitoring in operation and maintenance. *Automation in Construction*, 118: 103277.

Lu, Y., Peng, T. and Xu, X. (2019). Energy-efficient cyber-physical production network: Architecture and technologies. *Computers & Industrial Engineering*, 129: 56-66.

Madanayake, U.H. and Egbu, C. (2019). Critical analysis for Big Data studies in construction: Significant gaps in knowledge. *Built Environment Project and Asset Management*, 9(4): 530-547.

Mandolla, C., Petruzzelli, A.M., Percoco, G. and Urbinati, A. (2019). Building a Digital Twin for additive manufacturing through the exploitation of blockchain: A case analysis of the aircraft industry. *Computers in Industry*, 109: 134-152.

Mateev, M. (2020). Industry 4.0 and the Digital Twin for building industry. *Industry 4.0*, 5(1): 29-32.

Mawson, V.J. and Hughes, B.R. (2019). The development of modeling tools to improve energy efficiency in manufacturing processes and systems. *Journal of Manufacturing Systems*, 51: 95-105.

Min, Q., Lu, Y., Liu, Z., Su, C. and Wang, B. (2019). Machine learning-based Digital Twin framework for production optimization in petrochemical industry. *International Journal of Information Management*, 49: 502-519.

Mohammadi, N. and Taylor, J.E. (2017). Smart city Digital Twins. 2017 IEEE Symposium Series on Computational Intelligence (SSCI), IEEE.

Nasaruddin, A.N., Ito, T. and Tuan, T.B. (2018). Digital Twin approach to building information management. The Proceedings of Manufacturing Systems Division Conference 2018, The Japan Society of Mechanical Engineers.

Oliver, D., Adam, D. and Hudson-Smith, A. (2018). Living with a Digital Twin: Operational management and engagement using IoT and mixed realities at UCL's Here East Campus on the Queen Elizabeth Olympic Park. *Giscience and Remote Sensing*, GIS Research, UK (GISRUK).

Oracle (2017). *Digital Twins for IoT Applications: A Comprehensive Approach to Implementing IoT Digital Twins*. Oracle White Paper.

Pauwels, P., Zhang, S. and Lee, Y.-C. (2017). Semantic web technologies in AEC industry: A literature overview. *Automation in Construction*, 73: 145-165.

Press, G. Big Data definitions: What's yours. *Forbes Tech News*.

Qi, Q. and Tao, F. (2018). Digital twin and Big Data towards smart manufacturing and Industry 4.0: 360 degree comparison. *IEEE Access*, 6: 3585-3593.

Ruohomäki, T., Airaksinen, E., Huuska, P., Kesäniemi, O., Martikka, M. and Suomisto, J. (2018). Smart city platform enabling Digital Twin. 2018 International Conference on Intelligent Systems (IS), IEEE.

Salehi, S.A. and Yitmen, İ. (2018). Modeling and analysis of the impact of BIM-based field data capturing technologies on automated construction progress monitoring. *International Journal of Civil Engineering*, 16(12): 1-7.

Schleich, B., Anwer, N., Mathieu, L. and Wartzack, S. (2017). Shaping the Digital Twin for design and production engineering. *CIRP Annals*, 66(1): 141-144.

Shangguan, D., Chen, L. and Ding, J. (2019). A Hierarchical Digital Twin Model Framework for Dynamic Cyber-Physical System Design. Proceedings of the 5th International Conference on Mechatronics and Robotics Engineering.

Söderberg, R., Wärmefjord, K., Carlson, J.S. and Lindkvist, L. (2017). Toward a Digital Twin for real-time geometry assurance in individualized production. *CIRP Annals*, 66(1): 137-140.

Talkhestani, B.A., Jung, T., Lindemann, B., Sahlab, N., Jazdi, N., Schloegl, W. and Weyrich, M. (2019). An architecture of an intelligent Digital Twin in a cyber-physical production system. *at-Automatisierungstechnik*, 67(9): 762-782.

Tang, S., Shelden, D.R., Eastman, C.M., Pishdad-Bozorgi, P. and Gao, X. (2019). A review of building information modeling (BIM) and the internet of things (IoT) devices integration: Present status and future trends. *Automation in Construction*, 101: 127-139.

Tao, F., Qi, Q., Wang, L. and Nee, A. (2019). Digital Twins and cyber–physical systems toward smart manufacturing and Industry 4.0: Correlation and comparison. *Engineering*, 5(4): 653-661.

Tao, F., Sui, F., Liu, A., Qi, Q., Zhang, M., Song, B., Guo, Z., Lu, S.C.-Y. and Nee, A. (2019). Digital Twin-driven product design framework. *International Journal of Production Research*, 57(12): 3935-3953.

Tuegel, E.J., Ingraffea, A.R., Eason, T.G. and Spottswood, S.M. (2011). Reengineering aircraft structural life prediction using a Digital Twin. *International Journal of Aerospace Engineering*. 2011: 1-14. doi:10.1155/2011/154798

Wamba, S.F., Akter, S., Edwards, A., Chopin, G. and Gnanzou, D. (2015). How Big Data can make big impact: Findings from a systematic review and a longitudinal case study. *International Journal of Production Economics*, 165: 234-246.

Xu, J., Lu, W., Xue, F. and Chen, K. (2019). Cognitive facility management: Definition, system architecture, and example scenario. *Automation in Construction*, 107: 102922.

Xu, Y., Sun, Y., Liu, X. and Zheng, Y. (2019). A Digital-Twin-assisted fault diagnosis using deep transfer learning. *IEEE Access*, 7: 19990-19999.

Yang, C., Huang, Q., Li, Z., Liu, K. and Hu, F. (2017). Big Data and cloud computing: Innovation opportunities and challenges. *International Journal of Digital Earth*, 10(1): 13-53.

Yaqoob, I., Salah, K., Uddin, M., Jayaraman, R., Omar, M. and Imran, M. (2020). Blockchain for Digital Twins: Recent advances and future research challenges. *IEEE Network*.

Zhang, Y., Zhang, G., Wang, J., Sun, S., Si, S. and Yang, T. (2015). Real-time information capturing and integration framework of the internet of manufacturing things. *International Journal of Computer Integrated Manufacturing*, 28(8): 811-822.

Zhou, K., Fu, C. and Yang, S. (2016). Big Data driven smart energy management: From Big Data to big insights. *Renewable and Sustainable Energy Reviews*, 56: 215-225.

BIM-IoT-integrated Architectures as the Backbone of Cognitive Buildings: Current State and Future Directions

Ali Motamedi[*] and Mehrzad Shahinmoghadam

Département de génie de la construction, École de technologie supérieure, Université du Québec, Montréal, Québec

3.1 Introduction

Since the invention of computers, various research communities have been investigating computing systems that can naturally interact with the environment and humans. To achieve this vision, many significant research attempts have been made to enable computers to mimic the way human beings think, reason, and shape their behavior. Within the last few decades, the considerable developments in hardware and software engineering and rapid advancements in data mining and artificial intelligence (AI) has led to the emergence of the cognitive computing paradigm. Cognitive computing enables computer agents to reason and learn like the human brain does, namely, through the cycle of observation, interpretation, evaluation, and decision-making (Chen *et al.*, 2016). Although the subject of smart buildings is not new to built environment research, the notion of 'cognitive buildings' is a new line of research that investigates the potential of using the power of cognitive computing to make built environments more intelligent and human-centered.

Evidently, the availability of large amounts of historical data will be of critical importance in making it possible to extract new knowledge and improve the reasoning power of cognitive computing systems. Development of the Internet of Things (IoT) provides an immense

_____
*Corresponding author: Ali.motamedi@etsmtl.ca

opportunity to acquire valuable information about various features of buildings in real-time and to create large datasets for knowledge discovery. Due to rapid advancements in the domains of wireless sensing/actuation, embedded computing, and digital wireless communication technologies, the implementation cost and power consumption rate of IoT-enabled monitoring and control activities are declining. It is also interesting to note that building owners and facility managers are increasingly expressing interest in adopting IoT solutions. Building Information Modeling (BIM) is another existing opportunity which enables various aspects, disciplines, and systems of a building to be digitally constructed as a single digital model (Azhar Salman, 2011). By making building data available in a cloud-enabled BIM environment and merging the live streams of IoT data within it, digital models of buildings can be promptly updated as changes occur in the building and its surrounding environment. The availability of such BIM-IoT integrated frameworks provides an opportunity to create accurate Digital Twins of buildings. Finally, by adding cognitive computing power to such digital entities in the cloud, 'intelligent' and 'self-evolving' Digital Twins of buildings, or cognitive buildings can be created.

This chapter begins by defining cognitive buildings and showing how the use of the term 'cognitive' is different from other similar terms, such as 'smart' or 'intelligent'. Next, based on the provided definition, it is explained how IoT, BIM, and advanced analytics can be considered as three key technological disciplines that form the building blocks of cognitive buildings. Then, the main requirements relevant to the implementation of cognitive buildings along with their corresponding challenges and opportunities that address those challenges are discussed. A general architecture for cognitive buildings is presented. Finally, the chapter concludes by highlighting the available opportunities and future research directions.

3.2 Overview of Cognitive Buildings

3.2.1 Defining Cognitive Buildings

As mentioned earlier, the concept of 'smart building' is not new to the research community and the term is in fact extensively used in the industry, whereas 'cognitive building' is a relatively recent research theme. To understand the concept of cognitive buildings more clearly, the difference between a cognitive building and a smart building is discussed below.

Given that the definitions of *intelligence* and *cognition* and their interrelatedness are different within various fields of study, such as philosophy, neuroscience, and artificial intelligence (AI), this chapter adheres to AI's perspective to define intelligent and cognitive systems.

According to *Cambridge's Handbook of Artificial Intelligence* (Frankish and Ramsey, 2014), cognitive computing, which is among the newest research trends in the AI field, is primarily concerned with human-like intelligence through features, such as action selection, continuous learning, self-awareness, self-configuration, self-diagnostics, and self-healing. Systems with cognitive computing power will be able to autonomously make decisions for the required actions, operate in a complex and dynamic environment, be aware of their actions and the reasons for taking those actions, and adaptively cope with unanticipated circumstances. According to Chen *et al.* (2018), cognitive computing focuses on providing computer systems with brain-like intelligence that enables them to understand the objective world.

Thus, in general terms, a cognitive building can be defined as a building that has an intelligent and self-evolving Digital Twin which is able to effectively manage its operations and maintenance. In this respect, a technical definition for cognitive buildings could be: a building for which its lifecycle management is assisted by computerized systems, having cognitive computing power to enable features, such as situational/context-awareness, human-centered computing, self-evolution through continuous learning from experience, and fully/semi-automated decision making and actuation.

So, what makes cognitive buildings different from conventional smart buildings? The answer can be found in this analogy using the lighting management of a building as an example. A conventional smart building is considered as such when it is equipped with a Building Automation System (BAS), in this case for controlling the lighting level according to the data collected by the luminosity sensors installed within the building's various spaces. The BAS is provided with a knowledge base that contains predefined rules and a rule-based inference engine for decision-making. Although the building can be considered intelligent in terms of autonomously making decisions for lighting control, it lacks the ability to update its knowledge base according to the changing needs of the occupants. In contrast, a cognitive building is equipped with BIM/IoT systems and will be capable to learn new rules through interactions with the environment (including occupants) and autonomously updating its knowledge base throughout the building's lifecycle. Despite the significance of recent research on advanced building automation/managements systems (Tang *et al.*, 2020; Santos *et al.*,2020; Curry, 2020), the aforementioned analogy points to the need to shift from passive Building Management Systems (BMS) to systems that are equipped with human-like intelligence capabilities. Thus, researchers have started to pay attention to the benefits of cognitive computing for building/facility lifecycle management. For example, Xu *et al.* (2019) have most recently investigated the idea of 'cognitive facility management'. Their purpose has been to address the shortcomings of passive BMS in its ability to adapt to the complex and dynamic situations that are inherent to the realities of the objective world. Other research attempts relevant to the notion of cognitive building/facility can be found in Cheng *et al.* (2020); Desogus *et al.*(2017); Ibanez Stalin P. *et al.* (2019); Rinaldi *et al.* (2016). However, despite the growing research interest regarding the creation of cognitive buildings, the number of studies conducted in this regard is limited and major challenges remain. These challenges and the current opportunities that could help to address them are discussed later in this chapter.

3.2.2   Key Enablers of Cognitive Buildings

3.2.2.1 *Internet of Things/Cyber-Physical Systems*

Evidently automated collection and transmission of data from built environments play a major role in cognitive buildings. Previous research has proven the potential of Wireless Sensor Networks (WSNs) in this respect. However, rapid developments in remote sensing/actuation, wireless communications, and cloud computing have been unlocking the potential of Internet of Things (IoT), and Cyber-Physical Systems (CPS) to extensively collect live data from buildings. IoT holds the vision of vast networks of real-world objects equipped with embedded intelligence and sensing and actuation capabilities that are connected over the Internet (not Intranet or Extranet-based local networks), thereby enabling information flow and interactivity between virtual and physical worlds (Cirani *et al.*, 2018).

Although IoT shares close similarities with CPS and WSNs, Internet connectivity will be a distinguishing requirement for IoT applications (Minerva *et al.*, 2015). Hence, IoT makes it possible for a cognitive building to perform monitoring and control activities *anytime* and *anywhere*. In this light, IoT deployments are increasingly gaining importance as they are significant sources of data about buildings and their surrounding environment. As a result, the volume of research on the application of IoT in the operations and maintenance phase of the building's lifecycle has been expanding. Most recent examples of such research efforts can be found in Bedi *et al.* (2020); Cheng *et al.* (2020); Lu Qiuchen *et al.* (2020). Moreover, overviews of BIM and IoT-integrated frameworks have been recently reported in Shahinmoghadam and Motamedi (2019) and Tang *et al.* (2019). IoT sensors can collect data about the building's occupants' and/or operators' needs and expectations, as well as the monitoring data about the status of building components and environmental conditions. As the IoT data is transmitted to the middleware layer, intelligent computing agents can develop optimized action plans to be transmitted to the IoT actuators. This way, the building will be capable of changing its behavior to adapt to the dynamic conditions of its environment.

In order to increase their capacity to draw meaningful interpretations from raw IoT sensory data, researchers have started to investigate the potential of cognitive computing for IoT applications under the heading of Cognitive IoT (CIoT). A CIoT system is able to hypothesize correlations within IoT data and validate such hypotheses through evidence. By providing a better understanding of the changes occurring in the state of the physical elements, reflected within live IoT observations, CIoT facilitates situational-awareness about the physical world (Sheth, 2016). What makes an IoT application different from CIoT is the level of intelligence of each system. While detecting anomalies is a conventional and satisfactory IoT implementation application of machine learning, a combination of AI-based tools and techniques is needed to implement a CIoT system application. A CIoT system is able to solve problems that involve complex reasoning tasks well beyond the capacity of traditional machine learning. In the light of the vast potentials of CIoT to deliver higher levels of intelligence, researchers have recently shown interest in investigating such potentials in built-environment research (Hadj Sassi and Chaari Fourati, 2020; Zhang *et al.*, 2020).

3.2.2.2 Building Information Modeling and Digital Twinning

BIM processes significantly facilitate the digital representation and management of a building's various data/information. These data often originate from several fragmented disciplines and are continuously generated throughout the building's lifecycle. BIM environments facilitate the process of linking sensor data to other sources of building data (Desogus *et al.*, 2017). As mentioned earlier, merging rich building data contained in BIM models with live streams of IoT data in a cloud environment shapes a firm backbone for the creation of Digital Twins of buildings. However, in order to create such Digital Twins, several critical challenges must be addressed with regard to BIM-IoT integration, as recent studies reveal (Dave *et al.*, 2018; Tang *et al.*, 2019). By addressing such challenges and creating a BIM-IoT-enabled Digital Twin for a given building, large amounts of valuable data/information about the current and historical behaviors of the building will be organized in a single platform. In this way, the implementation of BMS will be significantly facilitated through integrated access to the linked building and sensor data that has been made available within the Digital Twins of buildings and that are being constantly updated with live streams of IoT data. Recent examples of the research works contributing to the creation of such Digital Twins have been reported in Al-Saeed *et al.* (2020), Alonso *et al.* (2019) and Boje *et al.* (2020).

General Electric (Boston, MA, USA), is one of the world's pioneers in Digital Twinning R&D. It views digital Twins according to a hierarchical structure that is composed of four levels: component, asset, system, and process (GE, 2020). If one were to generate a building's Digital Twin, that is, create a comprehensive Digital Twin of a building that contains the entire building's systems and processes using the currently available tools and methods, this would be extremely arduous, if not impossible. Hence, for the time being, for a given building it is expected that several Digital Twins will be deployed gradually and then be integrated within a single digital platform. In this light, since a cloud-enabled BIM environment provides a single web-based platform for an integrated representation and manipulation of building data, it has the significant potential to enable the gradual integration of individual Digital Twins that have been created over the course of a building's lifecycle. Taking, for example, the building's cooling/heating processes, detailed information, including sensory IoT data about the Digital Twins of the relevant components (e.g. valves and ducts), assets (e.g. cooling towers), and systems (e.g. the HVAC system), can be maintained within the BIM model as a unifying database in the cloud. Moreover, the actual cooling/heating processes and hypothetical scenarios can be replicated and visualized within the 4D-BIM simulation environments, as has been previously done (Natephra *et al.*, 2017b; Natephra and Motamedi, 2019).

3.2.2.3 Cognitive Computing

Inspired by the human mind's abilities, cognitive computing enables action planning through the integration and analysis of vast amounts of data coming from many disparate sources (Modha *et al.*, 2011). A cognitive computing system will be capable of reasoning based on the background knowledge, learning from experience, being aware of its own behavior, and appropriately responding to unexpected situations (Sheth, 2016). A review of the research publications on the subject of cognitive computing, and more specifically of publications with IoT/CIoT as the main theme, shows that there is confusion between cognitive computing and Big Data analytics. As indicated in the literature (Chen *et al.*, 2018), cognitive computing and Big Data are two distinct domains, both originating from the realm of data science. If the data that is being processed within a cognitive system complies with the main characteristics of Big Data, as has been highlighted in publications (Chen *et al.*, 2014), then it can be said that cognitive computing and Big Data analytics are working together. Accordingly, in this chapter, the notion of Big Data analytics is seen as a requirement for the creation of cognitive buildings (instead of as an enabler). A more detailed discussion in this regard is presented in the following section.

In a broader sense, cognitive computing can be performed based on two main approaches, namely, data-driven and knowledge-based driven computing systems. Data-driven approaches, which work on a data-mining basis, enable the extraction of new knowledge from large building datasets. Different methods of this approach have been previously investigated in the context of building lifecycle management. Examples of such studies include the application of Artificial Neural Networks (ANN) for energy consumption (Yuce *et al.*, 2014), indoor localization (Soltani *et al.*, 2015), and occupant temperature-preference learning (Peng *et al.*, 2019), Convolutional Neural Networks for textual building quality compliance data classification (Zhong *et al.*, 2019), Support Vector Machines (SVM) for energy efficiency (Shabunko *et al.*, 2014), association rule mining for building component fault diagnosis (Liu *et al.*, 2020), and a combination of ANN and SVM for building component predictive maintenance (Cheng *et al.*, 2020).

Knowledge-based computing approaches enable knowledge reuse by providing cognitive computing systems with logical reasoning capability. Examples of knowledge-based approaches used in built environment research include fault tree analysis for failure root-cause detection in facility maintenance (Motamedi *et al.*, 2014), case-based reasoning for building energy demand prediction (Monfet *et al.*, 2014), fuzzy inference systems for energy consumption prediction (Li and Su, 2010), and ontology-based systems for building environmental monitoring (Zhong *et al.*, 2018). To explain the role of knowledge representation in cognitive computing systems, Chen *et al.* (2018) put forth the following analogy: ordinary people and domain experts draw different interpretations from the same data, based on the depth of the background knowledge they possess in the relevant subject domain. The importance of providing background knowledge for cognitive buildings is more thoroughly discussed in the next section.

3.2.3 Hypothetical Scenario

To illustrate how the combination of IoT, BIM, and cognitive computing can lay a firm foundation for optimal cognitive buildings, a hypothetical scenario is examined as follows: A residential building located in a temperate-climate geographical zone, with extreme weather conditions (e.g. very cold and long winters, very hot and humid summers) has been equipped with an intelligent BMS combined with a BIM-IoT integrated framework. Thus, through this framework and the BIM models, detailed information about the building and its occupants' profiles is available and can be instantly accessed and updated in the cloud. With the help of numerous IoT sensors, the data about different variables, such as outdoor weather conditions, indoor environmental conditions, occupants' location, and the rate of energy consumption for each building component, are being recorded over a long period of time (e.g. four decades). Moreover, the way the building has been responding to specific environmental conditions, e.g. to maintain certain levels of satisfaction in terms of occupants' comfort and energy efficiency indicators, is being recorded. By applying the ANN method over the integrated sources of BIM and IoT data records, the building is capable of learning from past experiences. For instance, the predictive power of the trained ANN-based model can anticipate the required heating load and its distribution to meet a certain level of thermal comfort for a specific time-period during wintertime. Such predictions will be delivered, based on the most similar cases that have been previously experienced. The relevant information contained in the BIM database, such as the type of materials used in the building's structural envelope and their properties (e.g. thermal resistance), HVAC system specifications, occupant's profile, as well as the IoT data about the outdoor weather conditions, and indoor ambient monitoring data (e.g. air temperature and humidity), can shape a robust dataset, which can be used to assess the similarities between numerous conditions that the building has previously experienced. Consequently, the required working plan for the HVAC system can be recommended by evaluating the performance analysis results for the most similar previous cases. Obviously, a building operator's mind would be incapable of processing such a large dataset of detailed information about the building's energy performance accumulated over such a long period of time.

Now, for this residential building, over the last two years, due to abrupt changes in the region's climate, the pattern of weather conditions and the behavioral models of the building's occupants and their mental expectations have undergone recent changes. For example, the temperatures in this region have become higher than usual in the mid-winter days. In this case, the amount of data available to extract the optimized parameters for an efficient control of the HVAC system will be extremely limited. Although similar weather conditions can be found within the historical records, due to the occupants' emotional reactions to the unexpected changes in the regional climate, their sense of indoor thermal comfort might considerably differ for a given weather condition they are experiencing in the wintertime, compared to the same weather condition being experienced in the springtime. As a result, the limited size of available data for the new environmental conditions significantly impairs the performance of the trained ANN model. This is where knowledge-based systems can play an important role in preserving the cognitive computing power of the building. With the cognitive building's knowledge base and inference engine, logic-based interpretations can be drawn even if they are based on a limited size of input data. For example, the knowledge of experts in the field of building energy performance management can be formally represented in the form of IF-THEN rules with fuzzy variables (e.g. level of thermal comfort sensation), and the recommended course of action can be inferred through a fuzzy inference mechanism subsequently. The results of inferences then can be communicated to the IoT actuators to perform the physical action. Hence, a cognitive computing agent can address the data dependency issues that arise due to the unexpected climate changes and recommend an optimized course of action based on the experts' domain knowledge. However, eliciting knowledge from domain experts and providing formal representations of the acquired knowledge within computing platforms will be essentially challenging. Detailed discussion on the advantages and disadvantages of data-driven and knowledge-based approaches is given in the subsequent section.

3.3 Requirements, Challenges, and Opportunities

This section reviews a set of key requirements that are needed to create cognitive buildings using BIM-IoT integrated frameworks with embedded cognitive computing power. Moreover, the challenges impeding the fulfillment of these requirements, as well as the opportunities that are currently present and that could help to address them, are discussed below.

3.3.1 Human-centered Design

Human-centric design is used to deliver effective two-way interaction between humans and machines and is an imperative requirement for any smart environment/application. Two ways of ensuring the availability of proper human-machine interaction mechanisms are through communication channels and data/status visualization services and these have been most recently highlighted as two key requirements for developing Digital Twins with regards to buildings and cities (Lu Qiuchen *et al.*, 2020). The importance of developing cognitive buildings based on a human-centered architecture can be more clearly understood if one considers the following two main perspectives. First, a cognitive building needs to understand the changing needs of the occupants. Second, the current development of intelligent computing is still far from the vision of AI (i.e. a perfect replication of human cognition); thus, human intuition and feedback still play an indispensable role within the learning cycle of cognitive computing. The former refers to the need to provide smart environments with situational and context awareness, while the latter refers to the Human-In-The-Loop (HITL) notion.

With regards to the perspective that a cognitive building needs to understand the changing needs of its occupants, the key challenge is that an accurate understanding of occupant/user needs, particularly based on sensory data, is highly complicated due to the complex and dynamic nature of the environment (e.g. in the hypothetical scenario presented earlier, this refers to the occupant's unexpected behavior associated with the changing outdoor environmental conditions). One opportunity in this respect is to apply the recent advancements in context-aware (ubiquitous) computing for IoT/CPS systems. Contextual reasoning plays a significant role in drawing meaningful interpretations over raw sensor data, and context-aware computing has proven to be successful in this regard (Perera *et al.*, 2014). Although providing built environments with situational/contextual awareness is not a recent research trend, the interest of designing human-centered architectures according to the specific requirements of IoT/CIoT is relatively new. Recent studies (Jeon *et al.*, 2018; Pan *et al.*, 2019; Rashid *et al.*, 2019) show that researchers have already begun to grasp the potential of context-aware computing for IoT-enabled building systems. However, what is missing within the current literature are robust frameworks and methods for improving the effectiveness of context-aware computing by taking advantage of combining BIM and IoT data. Hence, further research needs to be done to narrow this knowledge gap. Another key opportunity is to apply reinforcement and deep learning algorithms to more profoundly understand the dynamic emotional states of occupants and more accurately recognize/predict their needs and behavioral patterns through natural language processing, speech recognition, and computer vision. An example of an application that directly relates to cognitive buildings can be found in Na *et al.* (2019) where a deep learning-based computer vision method facilitates the estimation of the occupants' metabolic rate to improve the accuracy of thermal comfort evaluations.

With respect to the second aspect (i.e. keeping human intuition in the loop), the main challenge is in providing effective interaction/communication mediums. For an HITL-driven CPS, conventionally dedicated interfaces in cyber environments (e.g. computer keyboards and joysticks) cannot effectively deliver the augmentation of human interactions (Schirner *et al.*, 2013). The rapid advancements in interactive software development, 3D graphics, and augmented/virtual reality (AR/VR) have provided vast opportunities in this respect by making it possible to deliver immersive and visually-rich experiences of cyber-physical interactivity. Hence, visual analytics systems, by connecting human analytical reasoning and data analytics through visualization and interactions with users (Cui, 2019), show great potential in maintaining the human intuition in the cyber-physical loop. Another such opportunity is that it is possible to expand the proven potential of both visual analytics and AR/VR-enabled interfaces with the help of BIM. Previous studies (Fukuda *et al.*, 2019; Motamedi *et al.*, 2014; Natephra *et al.*, 2017a; Natephra *et al.*, 2017b) have shown that a combination of BIM-based data representation/simulations and AR/VR-based visualizations can improve the design and maintenance of buildings by facilitating the incorporation of experts' heuristics for decision making and in-field problem solving. In the context of cognitive buildings, linking live streams of IoT data and rich building information contained in BIM databases to AR/VR-enabled interfaces can significantly facilitate the dissemination of the HITL paradigm. This will happen by facilitating the transmission of timely insights to the users and incorporating their intuitive feedback within the learning cycle. Examples of developing AR/VR-enabled interfaces for BIM-IoT integrated frameworks can be found in the literature (Carneiro *et al.*, 2018; Natephra and Motamedi, 2019; Patti *et al.*, 2017). Nevertheless, the number of such studies is considerably limited.

3.3.2 Interoperability

Creating cognitive buildings on top of BIM-IoT-enabled Digital Twins of buildings require an effective data/information exchange mechanism between BIM and IoT data ecosystems. However, data interoperability is a critical open research challenge for both BIM and IoT domains (Cirani *et al.*, 2018; Lemaire *et al.*, 2019; Shirowzhan *et al.*, 2020). This is mainly due to the fragmented and interdisciplinary nature of these domains, which has led to the abundance of numerous proprietary data-representation formats and schemas that are actively in use by various vendors. Hence, the severity of interoperability issues is further aggravated for BIM-IoT integration. Indeed, the shift from closed to open data ecosystems in BIM and IoT domains is an opportunity to deal with interoperability issues. Recently, Dave *et al.* (2018) presented a framework for BIM-IoT integration through open standards. However, considering the scale and complexity of the problem, the question of how open BIM standards (e.g. IFC, COBie) and IoT protocols can be linked to solve the interoperability problems remains to be answered more thoroughly and profoundly (Tang *et al.*, 2019). Other opportunities are currently available through latest developments in the fields of ontology engineering, semantic web (Berners-Lee *et al.*, 2001), and Linked Data (Bizer *et al.*, 2011). The tools and techniques from the aforementioned fields can be used to provide formal and semantically-rich descriptions of disparate and heterogeneous sources of BIM and IoT data, establishing semantic relationships between federated data sources (e.g. linking IoT sensory data to BIM spatial data), making such data understandable to both machines and humans, and enabling data exchange through open communication standards. Investigating the potential of ontologies and semantic web tools to address BIM-IoT interoperability though is an emerging line of research and more deliberate studies need to be conducted in this regard.

3.3.3   Big Data Analytics

Although the scale at which an IoT application is implemented can vary from small environment scenarios (e.g. smart home) to large environment scenarios (e.g. smart cities) (Minerva *et al.*, 2015), the IoT data generated within most implementation scenarios, including cognitive buildings scenarios, bear the main characteristics of Big Data. Such characteristics include high volume, high velocity, and high variety (Chen *et al.*, 2014). This said, as more systems become integrated within BIM environments (e.g. GIS, weather data, city models), the size and scope of BIM models eventually exceeds the capacity of the platforms that are currently in use to process BIM data. Hence, specialized Big Data processing platforms for BIM data will be needed in the near future (Bilal *et al.*, 2016; Peng *et al.*, 2017) and current research is investigating the efficient integration of Big Data analytics platform within BIM environments (Dou *et al.*, 2020; Lv *et al.*, 2020).

Recently, open challenges and best practices of applying Big Data analytics over IoT Big Data have been comprehensively discussed (Mahdavinejad *et al.*, 2018; Marjani *et al.*, 2017). However, with respect to BIM, there is a lack of such studies to identify the challenges and opportunities in the application of Big Data analytics over BIM Big Data. Hence, more in-depth research is needed to develop specialized Big Data analytics frameworks for cognitive buildings that take into account the specific characteristics of building BIM and IoT Big Data. The focus of such research efforts should be directed towards providing detailed guidance in the selection of the right type of analytics (e.g. off-line, real-time, memory-level), machine learning algorithms (e.g. classification, clustering, association analysis), and database technologies (e.g. object-oriented, graph-based), the right place for performing the analysis within the architecture (e.g. cloud, edge, fog), and the right existing technologies (e.g. Hadoop, Kafka, Spark).

3.3.4   Hybrid Analytics Architectures

A key requirement for cognitive buildings is that they should bear an acceptable degree of intelligence in order to draw meaningful inferences, regardless of the size of available data about the building and its environment. The challenge is that despite the significant potential of Big Data analytics approaches, the accuracy and reliability in their results greatly depends on the availability of the large volumes of training data. Hence, mere reliance on Big Data analysis will fall short in many prospective scenarios in which the size of historical data about the building and its environment is not sufficiently large. However, in contrast to data-driven analytics, knowledge-based computing agents are capable of delivering acceptable levels of accuracy and reliability when predicting the future states of a dynamic system, without having to process large volumes of datasets as the main reference for inference. Hence, the application of hybrid analytics architectures, which are composed of both data-driven and knowledge-based approaches, will offer an opportunity to address data-dependency issues.

The research community has been aware of the benefits of hybrid inference mechanisms in the development of smart built environments for some time. For example, to predict the rate of building energy consumption, the machine learning power of artificial neural networks has been combined with the case-based reasoning (Platon *et al.*, 2015) and traditional fuzzy inference systems in Li and Su (2010). More recent examples of using hybrid reasoning mechanisms that make use of a combination of historical data and domain knowledge to draw inferences can be found in Nguyen *et al.* (2020) and Sukor *et al.* (2019).

However, knowledge-based systems are subject to some critical limitations as well. One aspect of such limitations is related to the challenges of providing formal representations of expert knowledge. In particular, the degree to which human linguistic utterances can be represented in a machine-computable manner is still limited. This is, for the most part, due to the significant complexity of natural language, which plays a central role in knowledge externalization, and computers are at the early stages of understanding the profound semantics and pragmatics of the human language. Another key challenge stems from the fact that a significant portion of experiential knowledge has tacit traits. Tacit knowledge is highly personal and bound to a person's mental models, personal beliefs, and cognitive perspectives. As a result, it is vastly difficult to fully formalize and communicate tacit knowledge (Nonaka and Takeuchi, 1995), be it with human individuals or computing agents. Hence, the acquisition of the formal principles behind the building management experts' tacit (implicit) knowledge remains a challenging task in the creation of cognitive buildings. The application of formal ontologies along with context-aware computing principles, which have previously led to promising results (Kamsu Foguem *et al.*, 2008; Shahinmoghaddam *et al.*, 2018; Song *et al.*, 2016) for contextual knowledge representation and reasoning, provides an opportunity to deal with the inherent context-sensitivity characteristic of expert knowledge.

3.3.5 Robotic-based Assistance

The role of robots is an essential component of cyber-physical cognitive ecosystems (Chen *et al.*, 2018), and cognitive buildings in particular (Xu *et al.*, 2019). With respect to cognitive buildings, robots can significantly facilitate the automation of various building monitoring and control activities, either in a passive or proactive manner. In the passive scenario, robots receive action commands from the centralized or distributed cognitive computing agents, while in the proactive scenario, robots are capable of making decisions and taking actions independently through their embedded intelligence. For example, a mobile robot equipped with thermal and visual camera modules can be programmed to perform inspections and data gathering, and maintenance in an industrial plant based on received commands, or it can proactively monitor the environment through its own situational-awareness capability and plan the required actions. Most recent examples of providing robotic-based assistance to building operations and maintenance activities have been reported in Pakkala *et al.* (2020) and Krishna Lakshmanan *et al.* (2020).

Despite the immense potential robots offer to improve data collection and perform action commands for cognitive buildings, there are challenges to the integration of existing robotics systems within BIM-IoT-enabled Digital Twins of buildings. Such challenges mainly originate from two sources – first, lack of sufficient interoperability with BIM and IoT data within the robotic data ecosystems (Davtalab *et al.*, 2018; Vermesan *et al.*, 2017); second, the difficulty of providing robots with a profound understanding of BIM and IoT data. In other words, there exists a gap between human-centric and robot-centric perception of a given built environment (Turek *et al.*, 2017). As discussed earlier in this section, the application of formal ontologies offers considerable opportunities to address simultaneously both aspects of the aforementioned challenges. In particular, formal semantic descriptions provided within ontologies, not only facilitate interoperability through the use of shared vocabularies, but can also be used to draw logical inferences to establish semantic relationships between disparate sources of data (e.g. linking BIM geo-spatial data to the robot's IoT-enabled real-time location data), thereby providing a deeper understanding from low-level data.

3.3.6   Collective Intelligence

Allowing individual buildings to acquire new knowledge through a collective intelligence mechanism can be considered as another requirement for developing cognitive buildings. This will be similar to the way humans shape/adjust their behavior by observing different decisions/consequences by maintaining social interactions with others. Two key challenges in this respect will be to provide buildings with standard and open communication protocols for information exchange, and context-awareness to contextualize the information/knowledge that is being shared between them. The potential of formal ontologies and semantic web technologies in addressing these two challenges were discussed earlier in this section when reviewing interoperability and hybrid analytics architectures.

Additionally, the notion of 'swarm intelligence' offers a potential to equip BMS with an additional dimension of cognitive computing power. Swarm intelligence can be defined as a problem-solving paradigm in which individual agents within a family of information-processing units suggest solutions to a problem, refine those solutions through interactions with their peers, and finally select a solution, and adjust the variables according to their own needs (Kennedy, 2006). The shift from stand-alone intelligent agents working independently to a collaborative working environment in which those agents form a swarm, has been an ongoing strategic technological trend (CeArley *et al.*, 2017). Although swarm intelligence has been extensively investigated in robotics research, recently there has been an interest in investigating the potential of swarm intelligence and evolutionary algorithms for CPS with nature-inspired collective intelligence. For instance, a high-level architecture has been presented (Bagnato *et al.*, 2017), which aims at creating herds of heterogeneous CPS systems that will be capable of solving complex problems by demonstrating collective behaviors (e.g. surveillance of infrastructures through a swarm of unmanned aerial vehicles).

In built environment research, the notion of collective intelligence has been most recently investigated in the context of HVAC control systems (Yu and Li, 2020). The collective intelligence power of the proposed system was enabled by timely exchange of information about energy and flow-rate balance functions across neighboring nodes within distributed networks of HVAC systems. This led to more robust identification of flow-rates and pressure within the networks. In an earlier study, Chen and Hu (2018) highlighted that clustering buildings, to enable the open exchange of information and energy in recent studies, have shown promising results in different respects, such as environmental sustainability and energy cost saving. However, they have indicated that there is a knowledge gap regarding some of the key challenges in the clustering of buildings using swarm intelligence, namely, scalability and privacy issues. Challenges regarding the latter, i.e., privacy, along with security issues, are discussed below.

3.3.7   Security and Privacy

Evidently, similar to any information system, addressing security and privacy concerns will be a key requirement for cognitive buildings. The challenge of meeting security and privacy requirements for cognitive buildings, as BIM-centered IoT/CPS deployments, will be more problematic in at least two main aspects – first, when it comes to IoT and CPS systems (as is the case for cognitive buildings), the severity of cyber-security problems of CPS systems can be greater than that of conventional ICT systems, due to some major reasons as discussed in Greer *et al.* (2019), such as the possibility of cyber attacks targeting physical systems. In general, security and privacy issues have been marked as top-priority challenges to be addressed in IoT research (Al-Fuqaha *et al.*, 2015). Security challenges for IoT mainly stem from aspects, such as immaturity of standards, lack of secure design and development procedures for hardware/software, and inherent vulnerabilities of IoT devices compared to more robust endpoint devices, such as laptops and smartphones (Khan and Salah, 2018). Second, the known security challenges of cloud computing will put the successful adoption of cloud-enabled BIM at risk (Mahamadu *et al.*, 2013). Hence, considering the sensitivity of the data generated within built environments (e.g. the linkage with occupants' privacy and security), careful protective measures need to be adopted to mitigate the vulnerability of the less secure endpoints commonly used in buildings.

Blockchain technology was invented in 2008 as the key support for the crypto-currency bitcoin. Since the emergence of blockchain, academia and industry deemed it a key opportunity to help address IoT security issues (Khan and Salah, 2018; Reyna *et al.*, 2018). As a result, an increasing number of studies have been contributing to integrate blockchain into IoT/CPS systems. On the other hand, researchers have started to pay attention to the potential of blockchain for the built environment's cyber security, and particularly for cloud BIM. Examples of such studies include the work of Zheng *et al.* (2019) for securing the integrity, provenance, and traceability of BIM data in mobile computing, and the work of Elghaish *et al.* (2020) for securing financial transactions in the context of integrated project delivery systems. A recent review of blockchain integration with BIM processes has been reported (Nawari and Ravindran, 2019). Although the application of blockchain technology shows potential in addressing the security and privacy issues around BIM-IoT-integrated frameworks (Shahinmoghadam and Motamedi, 2019), more research is needed in the context of cognitive buildings that are built upon BIM-centered IoT/CPS deployments.

3.4 A Holistic BIM-IoT-Centered Architecture for Cognitive Buildings

A high-level architecture for the holistic design of cognitive buildings is presented in Fig. 3.1. The holistic approach taken herein provides the opportunity to meet the key requirements and address the multidisciplinary challenges discussed in the previous section.

The proposed architecture represents a cyber-physical ecosystem in which cognitive agents, including human individuals and cognitive computing units, can interact and communicate with each other and with

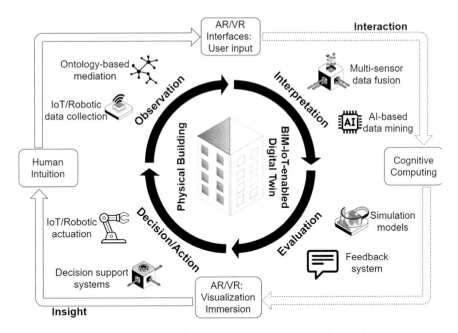

Fig. 3.1: High-level holistic architecture for cognitive buildings

the physical environment. As shown in Fig. 3.1, the proposed architecture places the BIM-IoT-enabled Digital Twin of the building at the center. Moreover, four stages have been considered to enable the evolution of the Digital Twin through the lifecycle of the building. It can be argued that these stages resemble the way human cognition evolves through a repetitive cycle of observation, interpretation/evaluation of the perceived data/information, and reasoning to making decisions and taking action.

In the observation stage, various types of sensors and robots collect data about the building, occupants, and the physical environment. IoT gateways designated for each local network of sensors, transmit the data to the BIM-centered middleware in the cloud. Once the data is transmitted to the cloud, an ontology-based mediation mechanism establishes semantic relationships between disparate sources of building data. This will be done with reference to the predefined semantic mappings between the local schemas of the data sources. Subsequently, an integrated access to federated sources of building data will be provided to conduct Big Data analytics in the interpretation stage. With an integrated access to the building's data in place, state-of-the-art approaches to multi-sensor data fusion, such as Dempster–Shafer and Bayesian theories, can be applied to relevant disparate sources of data to increase the reliability of the interpretations that will be drawn over single sources of sensory data. A

Big Data analytics framework that meets the specific requirements needed to deal with big BIM and IoT data (e.g. the right type of machine learning algorithms deployed on the right platform) can perform knowledge discovery using BIM-IoT-integrated data. In the evaluation stage, different simulation models, knowledge-based systems, as well as proper mechanisms able to receive user feedback must be put forward to evaluate the validity of the newly interpreted information/knowledge. Finally, in the decision/action stage, actuation commands will be delivered in a fully automated (i.e. through context-aware decision support systems) or semi-automated (i.e. through interventions by user-intuitive reasoning) manner. IoT actuators and robotic agents will execute actuation based on the actuation commands that have been conveyed through the IoT gateways. Within all stages, AR/VR interfaces will enable communication and interactivity with cognitive buildings through rich visualizations and immersive experiences, thereby enabling an HITL cognitive system.

3.5 Conclusion

This chapter began by reviewing the current literature to provide a clear definition for cognitive buildings and address some of the confusions around cognitive building, cognitive computing, and Big Data analytics. Using a hypothetical scenario as an example, it was shown that a combination of IoT/CPS, BIM and cognitive computing can lay a firm foundation in the creation of cognitive buildings. The key requirements needed to create cognitive buildings on the basis of BIM-IoT-enabled Digital Twins, as well as, the corresponding challenges and opportunities were discussed in this chapter. One of the main conclusions is that future BMS should be equipped with more human-centered interfaces, which enable visual analytics through rich visualization and interactive experiences. AR/VR-enabled interfaces provide immense opportunities in this regard, as alternatives to conventional user interfaces. Another conclusion is that further research must be conducted to enable the design of hybrid inference architectures that can compensate for the inherent shortcomings of data-driven and knowledge-based approaches used to confer buildings with cognitive computing power. The need to address the intrinsic vulnerability of IoT/CPS system, compared to the known security and privacy issues around conventional IT/ICT systems, is another key finding of this chapter. Another refers to the application of formal ontologies and semantic web technologies, which can be used to address a number of the relevant challenges, namely, interoperability, provision of shared understanding of data sources for different agents, and context-aware knowledge representation/reasoning, at the same time. However, it is worth noting that the development of new ontologies, in the context of cognitive buildings, does not appear to be a high priority for future research efforts. Alternatively, a promising direction for future research is the investigation of appropriate ontology matching/alignment approaches which can facilitate the reuse of the ontologies that already exist within the relevant domains (i.e. BIM, IoT, GIS, robotics). As open communications between various systems is of essential importance for the realization of cognitive buildings, standardization of data-representation schemas and exchange protocols, especially for BIM and IoT systems, will be a vital direction for future research. Finally, by resolving semantic heterogeneities that are a critical bottleneck for effective information/knowledge exchange between various cognitive facilities and enabling them with swarm intelligence, individual digital twins of various facilities (e.g. buildings, tunnels, bridges, and city infrastructures) could share insights in a real-time manner and act collectively, thereby forming smart cities.

References

Al-Fuqaha, A., Guizani, M., Mohammadi, M., Aledhari, M. and Ayyash, M. (2015). Internet of things: A survey on enabling technologies, protocols, and applications. *IEEE Communications Surveys Tutorials*, 17(4): 2347-2376.

Alonso, R., Borras, M., Koppelaar, R.H.E.M., Lodigiani, A., Loscos, E. and Yöntem, E. 2019. SPHERE: BIM Digital Twin Platform. *Proceedings*, 20(1): 9.

Al-Saeed, Y., Edwards, D.J. and Scaysbrook, S. (2020). Automating construction manufacturing procedures using BIM digital objects (BDOs): Case study of knowledge transfer partnership project in UK. *Construction Innovation*, 20(3): 345-377.

Azhar Salman (2011). Building Information Modeling (BIM): Trends, benefits, risks, and challenges for the AEC industry. *Leadership and Management in Engineering*, 11(3): 241-252.

Bagnato, A., Bíró, R.K., Bonino, D., Pastrone, C., Elmenreich, W., Reiners, R., Schranz, M. and Arnautovic, E. (2017). Designing swarms of cyber-physical systems: The H2020 CPSwarm Project: Invited Paper. *In*: Proceedings of the Computing Frontiers Conference, CF'17. Association for Computing Machinery, Siena, Italy, pp. 305-312.

Bedi, G., Venayagamoorthy, G.K. and Singh, R. (2020). Development of an IoT-driven building environment for prediction of electric energy consumption. *IEEE Internet of Things Journal*, 7(6): 4912-4921.

Berners-Lee, T., Hendler, J. and Lassila, O. (2001). The semantic web. *Scientific American*, 284(5): 28-37.

Bilal, M., Oyedele, L.O., Qadir, J., Munir, K., Ajayi, S.O., Akinade, O.O., Owolabi, H.A., Alaka, H.A. and Pasha, M. (2016). Big Data in the construction industry: A review of present status, opportunities, and future trends. *Advanced Engineering Informatics*, 30(3): 500-521.

Bizer, C., Heath, T. and Berners-Lee, T. (2011). Linked Data: The story so far.

Semantic Services, Interoperability and Web Applications: Emerging Concepts, 205-227.

Boje, C., Guerriero, A., Kubicki, S. and Rezgui, Y. (2020). Towards a semantic construction Digital Twin: Directions for future research. *Automation in Construction*, 114: 103179.

Carneiro, J., Rossetti, R.J.F., Silva, D.C. and Oliveira, E.C. (2018). BIM, GIS, IoT, and AR/VR integration for smart maintenance and management of road networks: A review. *In*: 2018 IEEE International Smart Cities Conference (ISC2), presented at the 2018 IEEE International Smart Cities Conference (ISC2), pp. 1-7.

CeArley, D., Burke, B., Searle, S. and Walker, M.J. (2017). *Top 10 Strategic Technology Trends for 2018*. Gartner, Inc. Orlando, Florida, United States.

Chen, M., Herrera, F. and Hwang, K. (2018). Cognitive computing: Architecture, technologies and intelligent applications. *IEEE Access*, 6: 19774-19783.

Chen, M., Mao, S. and Liu, Y. (2014). Big Data: A survey. *Mobile Netw Appl.*, 19(2): 171-209.

Chen, Y., Elenee Argentinis, J. and Weber, G. (2016). IBM Watson: How cognitive computing can be applied to Big Data challenges in life sciences research. *Clinical Therapeutics*, 38(4): 688-701.

Chen, Y. and Hu, M. (2018). A swarm intelligence-based distributed decision approach for transactive operation of networked building clusters. *Energy and Buildings*, 169: 172-184.

Cheng, J.C.P., Chen, W., Chen, K. and Wang, Q. (2020). Data-driven predictive maintenance planning framework for MEP components based on BIM and IoT using machine learning algorithms. *Automation in Construction*, 112: 103087.

Cirani, S., Ferrari, G., Picone, M. and Veltri, L. (2018). *Internet of Things: Architectures, Protocols and Standards*. John Wiley & Sons.

Cui, W. (2019). Visual analytics: A comprehensive overview. *IEEE Access*, 7: 81555-81573.

Curry, E. (2020). *Enabling Intelligent Systems, Applications, and Analytics for Smart Environments Using Real-time Linked Dataspaces*. Springer International Publishing, pp. 229-236.

Dave, B., Buda, A., Nurminen, A. and Främling, K. (2018). A framework for integrating BIM and IoT through open standards. *Automation in Construction*, 95: 35-45.

Davtalab, O., Kazemian, A. and Khoshnevis, B. (2018). Perspectives on a BIM-integrated software platform for robotic construction through Contour Crafting. *Automation in Construction*, 89: 13-23.

Desogus, G., Quaquero, E., Sanna, A., Gatto, G., Tagliabue, L.C., Rinaldi, S., Ciribini, A.L.C., Di Giuda, G. and Villa, V. (2017). Preliminary performance monitoring plan for energy retrofit: A cognitive building: The 'Mandolesi Pavillon' at the University of Cagliari. *In*: 2017 AEIT International Annual Conference, pp. 1-6.

Dou, S., Zhang, H., Zhao, Y., Wang, A., Xiong, Y. and Zuo, J. (2020). Research on construction of spatio-temporal data visualization platform for GIS and BIM Fusion. *The International Archives of Photogrammetry, Remote Sensing and Spatial Information Sciences*, Copernicus GmbH, 42: 555-563.

Elghaish, F., Abrishami, S. and Hosseini, M.R. (2020). Integrated project delivery with blockchain: An automated financial system. *Automation in Construction*, 114: 103182.

Frankish, K. and Ramsey, W.M. (2014). *The Cambridge Handbook of Artificial Intelligence*. Cambridge University Press.

Fukuda, T., Yokoi, K., Yabuki, N. and Motamedi, A. (2019). An indoor thermal environment design system for renovation using augmented reality. *Journal of Computational Design and Engineering*, 6(2): 179-188.

GE (2020). Digital Twin | GE Digital [WWW Document]. URL https://www.ge.com/digital/applications/digital-twin.

Greer, C., Burns, M.J., Wollman, D.A. and Griffor, E.R. (2019). *Cyber-Physical Systems and Internet of Things*. Pub n place

Hadj Sassi, M.S. and Chaari Fourati, L. (2020). Architecture for visualizing indoor air quality data with augmented reality-based cognitive internet of things. *In*: Advanced Information Networking and Applications, Advances in Intelligent Systems and Computing. Springer International Publishing.

Ibanez Stalin, P., Fitz, Theresa and Smarsly, Kay (2019). A semantic model for wireless sensor networks in cognitive buildings. Computing in Civil Engineering 2019, Proceedings, 234-241.

Jeon, Y., Cho, C., Seo, J., Kwon, K., Park, H., Oh, S. and Chung, I.-J. (2018). IoT-based occupancy detection system in indoor residential environments. *Building and Environment*, 132: 181-204.

Kamsu Foguem, B., Coudert, T., Béler, C. and Geneste, L. (2008). Knowledge formalization in experience feedback processes: An ontology-based approach. *Computers in Industry, Enterprise Integration and Interoperability in Manufacturing Systems*, 59(7): 694-710.

Kennedy, J. (2006). Swarm intelligence. *In*: *Handbook of Nature-Inspired and Innovative Computing*. Springer, pp. 187-219.

Khan, M.A. and Salah, K. (2018). IoT security: Review, blockchain solutions, and open challenges. *Future Generation Computer Systems*, 82: 395-411.

Krishna Lakshmanan, A., Elara Mohan, R., Ramalingam, B., Vu Le, A., Veerajagadeshwar, P., Tiwari, K. and Ilyas, M. (2020). Complete coverage path planning using reinforcement learning for Tetromino-based cleaning and maintenance robot. *Automation in Construction*, 112: 103078.

Lemaire, C., Rivest, L., Boton, C., Danjou, C., Braesch, C. and Nyffenegger, F. (2019). Analyzing BIM topics and clusters through ten years of scientific publications. *Journal of Information Technology in Construction (ITcon)*, 24(15): 273-298.

Li, K. and Su, H. (2010). Forecasting building energy consumption with hybrid genetic algorithm-hierarchical adaptive network-based fuzzy inference system. *Energy and Buildings*, 42(11): 2070-2076.

Liu, J., Shi, D., Li, G., Xie, Y., Li, K., Liu, B. and Ru, Z. (2020). Data-driven and association rule mining-based fault diagnosis and action mechanism analysis for building chillers. *Energy and Buildings*, 109957.

Lu Qiuchen, Parlikad Ajith Kumar, Woodall Philip, Don Ranasinghe Gishan, Xie Xiang, Liang Zhenglin, Konstantinou Eirini, Heaton James and Schooling Jennifer (2020). Developing a Digital Twin at building and city levels: Case study of West Cambridge Campus. *Journal of Management in Engineering*, American Society of Civil Engineers, 36(3): 05020004.

Lv, Z., Li, X., Lv, H. and Xiu, W. 2020. BIM Big Data Storage in WebVRGIS. *IEEE Transactions on Industrial Informatics*, 16(4): 2566-2573.

Mahamadu, A.-M., Mahdjoubi, L. and Booth, C. (2013). Challenges to BIM-cloud integration: Implication of security issues on secure collaboration. *In*: 2013 IEEE 5th International Conference on Cloud Computing Technology and Science, presented at the 2013 IEEE 5th International Conference on Cloud Computing Technology and Science, pp. 209-214.

Mahdavinejad, M.S., Rezvan, M., Barekatain, M., Adibi, P., Barnaghi, P. and Sheth, A.P. (2018). Machine learning for internet of things data analysis: A survey. *Digital Communications and Networks*, 4(3): 161-175.

Marjani, M., Nasaruddin, F., Gani, A., Karim, A., Hashem, I.A.T., Siddiqa, A. and Yaqoob, I. (2017). Big IoT data analytics: Architecture, opportunities, and open research challenges. *IEEE Access*, 5: 5247-5261.

Minerva, R., Biru, A. and Rotondi, D. (2015). Towards a definition of the Internet of Things (IoT). *IEEE Internet Initiative*, 1: 1-86.

Modha, D.S., Ananthanarayanan, R., Esser, S.K., Ndirango, A., Sherbondy, A.J. and Singh, R. (2011). Cognitive computing. *Communications of the ACM*, ACM, New York, NY, USA, 54(8): 62-71.

Monfet, D., Corsi, M., Choinière, D. and Arkhipova, E. (2014). Development of an energy prediction tool for commercial buildings using case-based reasoning. *Energy and Buildings*, 81: 152-160.

Motamedi, A., Hammad, A. and Asen, Y. (2014). Knowledge-assisted BIM-based visual analytics for failure root cause detection in facilities management. *Automation in Construction*, 43: 73-83.

Na, H., Choi, J.-H., Kim, H. and Kim, T. (2019). Development of a human metabolic rate prediction model based on the use of Kinect-camera generated visual data-driven approaches. *Building and Environment*, 160: 106216.

Nguyen, D., Nguyen, L. and Nguyen, S. (2020). A novel approach of ontology-based activity segmentation and recognition using pattern discovery in multi-resident homes. *In*: Frontiers in Intelligent Computing: Theory and Applications, Advances in Intelligent Systems and Computing. Springer, Singapore, pp. 167-178.

Natephra, W. and Motamedi, A. (2019). Live data visualization of IoT sensors using augmented reality (AR) and BIM. *In*: 2019 Proceedings of the 36th ISARC, Banff, Alberta, Canada, pp. 632-638.

Natephra, W., Motamedi, A., Fukuda, T. and Yabuki, N. (2017a). Integrating building information modeling and virtual reality development engines for building indoor lighting design. *Visualization in Engineering*, 5(1): 19.

Natephra, W., Motamedi, A., Yabuki, N. and Fukuda, T. (2017b). Integrating 4D thermal information with BIM for building envelope thermal performance analysis and thermal comfort evaluation in naturally ventilated environments. *Building and Environment*, 124: 194-208.

Nawari, N.O. and Ravindran, S. (2019). Blockchain technology and BIM process: Review and potential applications. *Journal of Information Technology in Construction (ITcon)*, 24(12): 209-238.

Nonaka, I. and Takeuchi, H. (1995). *The Knowledge-creating Company: How Japanese Companies Create the Dynamics of Innovation*. Oxford University Press.

Pakkala, D., Koivusaari, J., Pääkkönen, P. and Spohrer, J. (2020). An experimental case study on edge computing-based cyber-physical digital service provisioning with mobile robotics. pp. 1165-1174. *In*: Proceedings of the 53rd

Hawaii International Conference on System Sciences. Honolulu County, Hawaii, United States.

Pan, Z., Hariri, S. and Pacheco, J. (2019). Context-aware intrusion detection for building automation systems. *Computers & Security*, 85: 181-201.

Patti, E., Mollame, A., Erba, D., Dalmasso, D., Osello, A., Macii, E. and Acquaviva, A. (2017). Information modeling for virtual and augmented reality, *IT Professional*, 19(3): 52-60.

Peng, Y., Lin, J.-R., Zhang, J.-P. and Hu, Z.-Z. (2017). A hybrid data mining approach on BIM-based building operation and maintenance. *Building and Environment*, 126: 483-495.

Peng, Y., Nagy, Z. and Schlüter, A. (2019). Temperature-preference learning with neural networks for occupant-centric building indoor climate controls. *Building and Environment*, 154: 296-308.

Perera, C., Zaslavsky, A., Christen, P. and Georgakopoulos, D. (2014). Context-aware computing for the Internet of Things: A survey. *IEEE Communications Surveys Tutorials*, 16(1): 414-454.

Platon, R., Dehkordi, V.R. and Martel, J. (2015). Hourly prediction of a building's electricity consumption using case-based reasoning, artificial neural networks and principal component analysis. *Energy and Buildings*, 92: 10-18.

Rashid, K.M., Louis, J. and Fiawoyife, K.K. (2019). Wireless electric appliance control for smart buildings using indoor location tracking and BIM-based virtual environments. *Automation in Construction*, 101: 48-58.

Reyna, A., Martín, C., Chen, J., Soler, E. and Díaz, M. (2018). On blockchain and its integration with IoT: Challenges and opportunities. *Future Generation Computer Systems*, 88: 173-190.

Rinaldi, S., Bittenbinder, F., Liu, C., Bellagente, P., Tagliabue, L.C. and Ciribini, A.L.C. (2016). Bi-directional interactions between users and cognitive buildings by means of smartphone app. Presented at the 2016 IEEE International Smart Cities Conference (ISC2), IEEE, pp. 1-6.

Santos, G., Vale, Z., Faria, P. and Gomes, L. (2020). BRICKS: Building's reasoning for intelligent control knowledge-based system. *Sustainable Cities and Society*, 52: 101832.

Schirner, G., Erdogmus, D., Chowdhury, K. and Padir, T. (2013). The future of human-in-the-loop cyber-physical systems. *Computer*, 46(1): 36-45.

Shabunko, V., Lim, C.M., Brahim, S. and Mathew, S. (2014). Developing building benchmarking for Brunei Darussalam. *Energy and Buildings*, 85: 79-85.

Shahinmoghadam, M. and Motamedi, A. (2019). Review of BIM-centered IoT deployment: State of the art, opportunities, and challenges. *In*: 2019 Proceedings of the 36th ISARC, Banff, Alberta, Canada, pp. 1268-1275.

Shahinmoghaddam, M., Nazari, A. and Zandieh, M. (2018). CA-FCM: Towards a formal representation of expert's causal judgements over construction project changes. *Advanced Engineering Informatics*, 38: 620-638.

Sheth, A. (2016). Internet of Things to Smart IoT through semantic, cognitive, and perceptual computing. *IEEE Intelligent Systems*, 31(2): 108-112.

Shirowzhan, S., Sepasgozar, S.M.E., Edwards, D.J., Li, H. and Wang, C. (2020). BIM compatibility and its differentiation with interoperability challenges as an innovation factor. *Automation in Construction*, 112: 103086.

Soltani, M.M., Motamedi, A. and Hammad, A. (2015). Enhancing cluster-based RFID tag localization using artificial neural networks and virtual reference tags. *Automation in Construction*, 54: 93-105.

Song, B., Jiang, Z. and Liu, L. (2016). Automated experiential engineering knowledge acquisition through Q&A contextualization and transformation. *Advanced Engineering Informatics*, 30(3): 467-480.

Sukor, A.S.A., Zakaria, A., Rahim, N.A., Kamarudin, L.M., Setchi, R. and Nishizaki, H. (2019). A hybrid approach of knowledge-driven and data-driven reasoning for activity recognition in smart homes. *Journal of Intelligent & Fuzzy Systems*, IOS Press, 36(5): 4177-4188.

Tang, S., Shelden, D.R., Eastman, C.M., Pishdad-Bozorgi, P. and Gao, X. (2019). A review of building information modeling (BIM) and the internet of things (IoT) devices integration: Present status and future trends. *Automation in Construction*, 101: 127-139.

Tang, S., Shelden, D.R., Eastman, C.M., Pishdad-Bozorgi, P. and Gao, X. (2020). BIM-assisted building automation system information exchange using BACnet and IFC. *Automation in Construction*, 110: 103049.

Turek, W., Cetnarowicz, K. and Borkowski, A. (2017). On human-centric and robot-centric perspective of a building model. *Automation in Construction*, 81: 2-16.

Vermesan, O., Bröring, A., Tragos, E., Serrano, M., Bacciu, D., Chessa, S., Gallicchio, C., Micheli, A., Dragone, M. and Saffiotti, A. (2017). *Internet of Robotic Things: Converging Sensing/Actuating, Hypoconnectivity, Artificial Intelligence and IoT Platforms*. River Publishers.

Xu, J., Lu, W., Xue, F. and Chen, K. (2019). Cognitive facility management: Definition, system architecture, and example scenario. *Automation in Construction*, 107: 102922.

Yu, Z. and Li, H. (2020). Identification of flow-rates and pressures in HVAC distribution network based on collective intelligence system. *In*: Proceedings of the 11th International Conference on Modelling, Identification and Control (ICMIC2019). Lecture Notes in Electrical Engineering, Springer, Singapore, pp. 1229-1237.

Yuce, B., Li, H., Rezgui, Y., Petri, I., Jayan, B. and Yang, C. (2014). Utilizing artificial neural network to predict energy consumption and thermal comfort level: An indoor swimming pool case study. *Energy and Buildings*, 80: 45-56.

Zhang, Y., Tian, G. and Chen, H. (2020). Exploring the cognitive process for service task in smart home: A robot service mechanism. *Future Generation Computer Systems*, 102: 588-602.

Zheng, R., Jiang, J., Hao, X., Ren, W., Xiong, F. and Ren, Y. (2019). bcBIM: A Blockchain-based Big Data model for BIM modification audit and provenance in mobile cloud. *Mathematical Problems in Engineering*, Hindawi.

Zhong, B., Gan, C., Luo, H. and Xing, X. (2018). Ontology-based framework for building environmental monitoring and compliance checking under BIM environment. *Building and Environment*, 141: 127-142.

Zhong, B., Xing, X., Love, P., Wang, X. and Luo, H. (2019). Convolutional neural network: Deep learning-based classification of building quality problems. *Advanced Engineering Informatics*, 40: 46-57.

The Evolution of Building Information Model: Cognitive Technologies Integration for Digital Twin Procreation

Gözde Basak Özturk

Faculty of Engineering, Civil Engineering Department, Aydin Adnan
Menderes University, Central Campus, P. Box: 09010 Aydın, Turkey

4.1 Introduction

Considering integration of Building Information Modeling (BIM) and
Internet of Things (IoT), there are numerous recent developments, not
only focus on implementation and research, but also put great efforts in
improving the performance from the perspective of management and
organization. These improvements can be classified as technology-related
issues and information-related issues. The effectiveness of architecture,
engineering, construction, operations, and facility management (AECO/
FM) activities heavily relies on continuous information acquisition,
share, store, and use and reliable communication channels and properly
integrated real-time data. BIM can aid in overcoming some of the complex
problems in the AECO/FM activities. However, the performance of BIM-
enabled projects can be increased via integration with available digital
and cognitive technologies.

Digital Twin implementation started approximately a decade ago
with the proposal for next generation fighter aircraft and NASA vehicles.
Aerospace industry implementations became a flagship for Internet of
Things (IoT) integration in relation with product lifecycle management

Email: gbozturk@adu.edu.tr

in Industry 4.0. However, the primary use of the Digital Twin concept is in engineered products, production machines or production lines. The concept of implementation in BIM methodology evolution to achieve BIM Digital Twins is a challenge for the AECO/FM industry. Nowadays, there is a lot of information from the building and its subsystems. The integration with IoT may provide opportunity to capture, store, and share critical building information. The integrated BIM platform will facilitate improvements in the AECO/FM activities. In addition, it will improve efficiency and effectiveness throughout the project lifecycle due to a seamless integration of each dimension, IoT and respective stakeholders within the platform. The solution will integrate two planes of research, innovation, and improvement through an integration of the processes under the Digital Twin concept involving architecture, engineering, construction, operation, and facility management processes based on large-scale data, information and knowledge integration, and synchronization, thus enabling better handling and processing.

4.2 Building Information Modeling (BIM)

Building Information Modeling (BIM) is a technology that serves as a flexible platform open to integrating nD data and other technologies to create, share, and use information throughout the project and building lifecycle. BIM has become the pioneering method for integrated delivery with its promise of digitalization of the built environment. BIM model is a digital representation of its physical counterpart as a building and is integrated with multi-dimensional information as a knowledge creation and sharing tool for building information. Ozturk (2020) and Juan *et al.* (2017) state that BIM has gained considerable recognition by reducing time and cost and improving the quality of the construction project. Being not a common practice, its implementation is becoming widespread in large and complex projects for its ability to create more efficient projects for reduced claims, clear and fast analysis, multi-purpose serving visualizations, and increased collaboration among all stakeholders (Hwang *et al.*, 2019; Chen and Tang, 2019; Sanchez and Joske, 2016). BIM benefits all the stakeholders while ensuring the scope, improving performance, reducing the number of change and financial risk with reliable estimates, improving marketing, improving the design, better safety, improving communication, supporting decision making, integrating infinite data, sharing reliable information, and many more.

Researchers state that BIM is a digital format on which project design and building information is built and used throughout the project lifecycle (Tsilimantou *et al.*, 2020; Chang *et al.* 2018; Sanchez and Joske, 2016; Shou *et al.*, 2015). The use of this information requires guidelines and procedures. Integrating many dimensions of building information on the platform requires the use of multiple softwares. Each dimension has its own merit (Fig. 4.1). While second and third dimensions require shape and size of spatial elements, fourth and fifth dimensions require the time and cost information mostly used during the construction process, where the sixth dimension requires energy-related information, seventh dimension represents the facility management-related information on the BIM model.

Interoperability enables data transfer from one software to another by facilitating workflow and promoting automation. There are several suppliers that offer software products with various specifications for different specialities within the project. Interoperability requires all suppliers to use an Industry Foundation Class data model (IFC), which is a neutral data model that allows the use of all suppliers and basic exchange of information about the project, regardless of their area of expertise (Ozturk, 2020; Papadonikolaki *et al.*, 2019; Lai *et al.*, 2019; Solibri, 2016).

4.2.1 BIM Capability Stages

BIM capability is determined as the basic ability to perform a task via BIM method or deliver a service/product with BIM technology (Mahamadu *et al.*, 2019; Yilmaz *et al.*, 2019). BIM stages represent minimum BIM requirements or important milestones that must be achieved by project teams or organizations during BIM implementation (Kassem and Succar, 2017; Ahmed and Kassem, 2018) (Fig. 4.2).

BIM Stage 1: Object-based modeling
BIM Stage 2: Model-based collaboration
BIM Stage 3: Network-based integration

BIM stages are determined by base requirements. BIM Capability Stage 1 requires an association to deliver object-based modeling software. BIM Capability Stage 2 requires an association to collaborate through a multidisciplinary approach based on a model in a project. BIM Capability Stage 3 requires an association to share object-based models through a network-based method with other various disciplines.

4.2.2 BIM Maturity Levels – Integration Levels

The term 'BIM Maturity' implies to the repeatability, quality, and level of greatness inside a BIM Capability (Wu *et al.*, 2017; Smits *et al.*, 2016; Liang *et al.*, 2016). BIM Maturity's criteria are execution improvement levels that firms and associations seek to or move in the direction of. As a rule, the advancement from low to more high levels demonstrates better control through limiting varieties between execution targets and genuine outcomes, better consistency and determining by bringing down changeability in competency, execution and costs, and more noteworthy

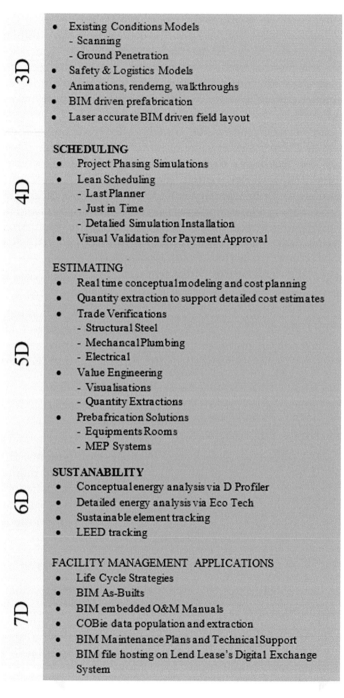

Fig. 4.1: Dimensions of BIM

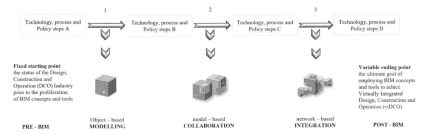

Fig. 4.2: BIM capability stages (adapted from Succar *et al.*, 2013)

viability in arriving at characterized objectives and setting new increasing yearning ones.

The idea of BIM Maturity has been received from Software Engineering Institute's (SEI) Capability Maturity Model – a procedure improvement structure at first proposed as an apparatus to assess the capacity of government's temporary workers to convey a product venture. In general, maturity models are composed of more than one maturity level (Kassem and Succar, 2017; Siebelink *et al.*, 2018; Ahankoob *et al.*, 2018; Azzouz and Hill, 2017). At the point when the necessities of each level are fulfilled, implementers would then be able to expand on the set up parts to endeavor 'higher' maturity. Research led inside different ventures has just distinguished the connection between improving process maturity and business productivity. In spite of the fact that there are different endeavors received from CMM, which center around the construction industry, there is no thorough maturity model/index that can be applied to BIM, its usage stages, players, expectations, or its impact on project lifecycle stages. To overcome this limitation, the BIM Maturity Index (BIMMI) has been created by investigating and afterward coordinating a few development models utilized across various industries (Lu *et al.*, 2018; Kassem and Succar, 2017). It has been tweaked to mirror the points of interest in BIM capability, productivity targets, execution requirements, and quality management. The BIM Maturity Index can be categorized under five levels:

a. Initial/Ad-hoc
b. Defined
c. Managed
d. Integrated and
e. Optimized

4.3 Internet of Things (IoT)

The Internet of Things (IoT) concept was first initiated in the 1980s right after the emergence of the Internet. Thereafter, the idea of connecting appliances came up (Raji, 1994). Radio-frequency identification (RFID) and IoT integration constructed the main structure of this concept. The research tendency in the Internet of Things (IoT) subject has considerably aroused in recent years. As presented in Fig. 4.6, the number of research documents about IoT research in architecture, engineering, construction, operations, and facility management (AECO/FM) industry has risen considerably, starting from 2015 to date. The improvements in IoT research results in integrated studies of IoT and UWB, WLAN, artificial intelligence, machine learning, and iBeacon technologies (Cheng *et al.*, 2020; Kong and Ma, 2020; Costin *et al.*, 2019; Allam and Dhunny, 2019; Tang *et al.*, 2019; Wigmore, 2014). IoT is the communication, computation, and coordination among machines and objects via the Internet. IoT is described as a dynamic network of data collectors, sensors, computers, and machines erected on a platform of telecommunication protocols (Internet), interact with each other to analyze, process, and store data in a cloud-based server for facilitating some services and processes (Niu *et al.*, 2019, Joel *et al.*, 2019; Carmona *et al.*, 2018; Lee and Lee, 2015). The IoT devices must have a permanent connection to a platform for a robust interaction among each other (Košťál *et al.*, 2019; He *et al.*, 2014). The Internet of Things concept can better be internalized by connecting different paradigms. An IoT platform consists of layers of sensors, processors, gateways, and applications (Sung *et al.*, 2019; Zhao *et al.*, 2019; Gamil *et al.*, 2020; Opentechdiary, 2015; He *et al.*, 2014). These sensors are attached to machines and objects to collect real-world data from the physical environment. The information is eliminated from unnecessary data and stored in local storage from where the information is transferred to cloud storage. The collective information from many objects and machines is stored in cloud storage and used to manage and control the system. Radio Frequency Identification (RFID) (Zhang *et al.*, 2016), wireless sensor network (WSN) (Whitmore *et al.*, 2015), and near-field communication (NFC) (Nagashree *et al.*, 2014), ZigBee (Wang *et al.*, 2011; Zillner, 2018; Peng and Huang 2016, Talaviya *et al.*, 2013; Salleh *et al.*, 2013), Bluetooth, Wi-Fi, 3G-4G-5G Network, and other communication technologies are utilized for IoT implementation. IoT is mainly utilized in supply-chain management, delivery, security subjects in healthcare, smart home, animal tracking, smart robotics, smart transportation, infrastructure management, manufacturing, smart building, smart agriculture, smart retail, and in other areas.

The benefits of using IoT technology eases and simplifies to automate, achieve, and control tasks (Häikiö *et al.*, 2020; Costin *et al.*, 2019). However, security and privacy are the main considerations in IoT implementation (Noor and Hassan, 2020; Atlam and Wills, 2020). The IoT elements needed for a functional implementation are namely identification, communication, sensing, services, computation, and semantics. The IoT architecture, that is composed of three layers in general, consists of the perception layer, network layer, and application layer (Fig. 4.3). Despite the fact that four-layered (perception, support, network, and application layers) (Fig. 4.4), five-layered (perception, transport, processing, application, and business layers) (Fig. 4.4), and six-layered (perception, observer, processing,

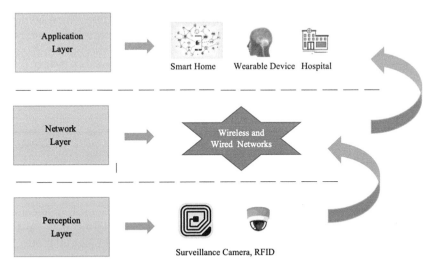

Fig. 4.3: The three-layered architecture of IoT

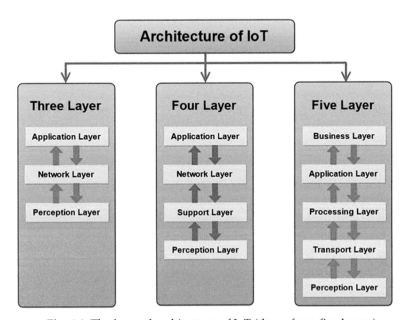

Fig. 4.4: The layered architecture of IoT (three, four, five layers)

security, network, and application) (Fig. 4.5) IoT architecture types were proposed since security and privacy is of vital concern.

The perception layer is a sensor layer that functions as the perceptive organs of a human in identifying objects and collecting information responsibilities. The common security concerns are mostly related to sensors. The network layer works as a transmission layer which bridges the perception layer and the application layer. The collected information is transmitted by wireless or wire-based systems via the network layer to the application layer. The responsibility of the network layer is to connect all smart objects, network devices, and networks. The application layer defines all applications using the IoT technology or in which IoT is deployed (Burhan *et al.*, 2018). The services depending on the information need of the application are provided to applications, like smart homes, smart cities, smart transportation, and others by the application layer. In the four-layered IoT architecture, there is one more layer that differs from the three-layered IoT architecture called the support layer (Fig. 4.4). This layer is designed to prevent attacks and achieve more security. The support layer becomes a buffer zone between network and application layers with two responsibilities of verifying the authentic user as the sender of the information and protecting information against attacks

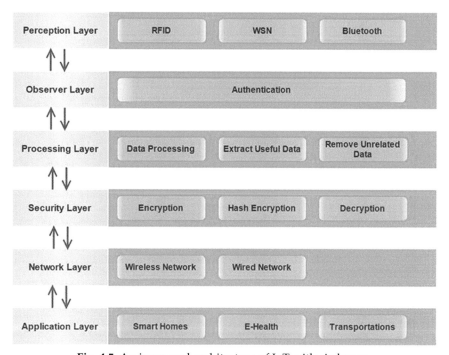

Fig. 4.5: An improved architecture of IoT with six layers

and sending information to the application layer. The five-layered IoT architecture consists of processing and business layers different from the previous systems (Fig. 4.4). Information sent from a transport layer is collected within the processing (middleware) layer for eliminating the unnecessary information and consolidating the useful information. The business layer manages and controls the intended action of an application for the whole system. This layer determines the method for information creation, storage, and change.

The six-layered IoT architecture proposed by Burhan *et al.* (2018) consists of extra layers that differ from other IoT architecture systems that are observer and security layers (Fig. 4.5). Observer (monitor) layer receives information from the perception layer to check whether the information is secured from viruses and intruders or not. If there is any suspicious situation, this layer blocks the information transmission. The authentication is also checked in this layer. Since the network layer is attacked for getting information from the users, the security layer is used as the protector of the subsequent layer of the network. This protection is achieved via use of keys for encryption of the information to be sent in cipher-text format to the network layer. The authenticated user has the key to decrypt the chipper text into a readable original version.

IoT uses standard protocols and technologies. Technologies, such as sensing, identification, recognition, positioning technologies; hardware, software, cloud and network platforms; and power and energy storages, security systems can be enabled via IoT. Information-sharing devices sense and actuate by interconnection across platforms. Integration of BIM and IoT devices provides a robust sample model for applications to increase efficiency in operation and construction. Integrating the rapidly-evolving IoT technology into BIM models enables defining of the designed three-dimensional objects, features, and spatial organization as a virtual asset set and creating a high-quality data set (Zhou, 2012). An activity-sensitive object can record information about business activities and their use. A policy-sensitive object is an activity-sensitive object that can interpret events and activities according to predefined organizational policies. They can simply be characterized in three ways: these are awareness, representation and interaction. An aware object understands the world in terms of event and activity streams, the application model consists of aggregation functions to accumulate activity over time, and activity-sensitive objects prioritize saving data and do not provide interactive capabilities. Depending on the temperature, vibrations, and relative proximity of objects, IoT-based devices can inform employees about security breaches and prod them to take appropriate action (Kortuem *et al.*, 2010).

4.3.1 IoT and BIM Applications

With IoT, the need for rich metadata and interconnection can be solved. There are too many connections between all parts of the project to become real for innovative works. IoT has many options to make this connection issue solved. The connection of the designed and built-environments problem is solved with the technology behind IoT. Interconnections among many parts of the project will be easier with the fast network and the team is also included in these parts. Cloud platforms play an important role in the interconnection options. Advanced selections become real with growing cloud technologies. Semantic web-based systems are also another supporter of this interconnection. Before smart devices, the market could not realize this much data transfer but with a growing sector of smart devices, these cloud, semantic web, automation, etc. technologies have developed too. This development becomes a necessity for IoT with the need for smart houses, smart cars, and automation systems likely to increase.

The integration of IoT devices and the BIM model with real-time data can improve efficiency in the processes of construction projects (Cheng *et al.*, 2020; Edirisinghe and Woo, 2020; Puyan *et al.*, 2019; Rogage *et al.*, 2019; Tang *et al.*, 2019). There are numerous application areas of real-time data streaming to the BIM model via IoT devices. IoT sensor networks provide complementary data with BIM models for a broader view of the construction project throughout the lifecycle. BIM model offers geometrical and spatial designed building-information dataset composed of virtual assets that can be operated. IoT-sensor networks provide real-time data among IoT devices to enhance improvement throughout the processes of the construction project. Mechanisms, such as programming APIs associated with these applications, manual interfaces of proprietary systems, export via open standards, and potential database connections to the systems enable access to BIM and IoT data (Wu and Liu, 2020; Zhai *et al.*, 2019; Quinn *et al.*, 2020). However, research on IoT and BIM integration is still not intensively focused yet (Tang *et al.*, 2019).

4.3.2 IoT Use in the AECO/FM Industry

The AECO/FM industry recently had a very intensive tendency on IoT (Fig. 4.6). The number of articles published between 2006 and 2019 is displayed yearwise in Fig. 4.6. The number of papers related to IoT for AECO/FM has gone up considerably in this period. Especially in 2016, the number of papers displays a critical upturn. The average number of papers in the last five years has been around 225. The trend of the research in IoT for the AECO/FM industry shows exponential growth, which highlights a major impact on the community.

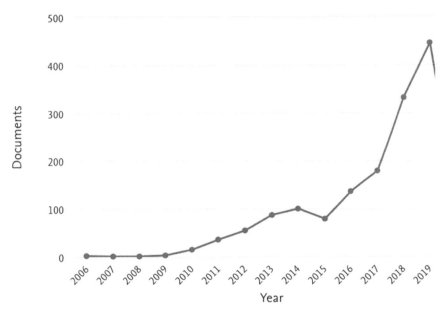

Fig. 4.6: Yearly number of articles from 2006 to 2019 on IoT research for the AECO/FM industry (the article numbers in 2020 are not shown since publications have not been completed yet)

The result of the co-occurrence analysis of keywords is mapped and illustrated with a network diagram in Fig. 4.7. The color of a keyword determines the cluster to which the keyword belongs. The keywords were clustered by colors in five categories (Fig. 4.7). The clusters were named after the field of study that the keyword belonged to: *quality management and monitoring by information management, lifecycle management by BIM integration, cognitive technology integration, Big Data management,* and *automation and energy management.*

The keywords are represented in circles and labels (Fig. 4.7). The size of the circle of an item is determined by the weight of the item. The higher weight of a keyword is represented by a larger size of the circles and labels. The distance between two keywords in the visualization indicates approximately the relatedness of the keywords in terms of co-occurrence links. In general, the closer two keywords are located to each other, the stronger is their relatedness.

The research in IoT for the AECO/FM industry mostly focused on *Big Data management and cognitive technology integration* for IoT technology adoption in the industry activities. The research output of IoT technology-adoption studies mostly focused on *automation and energy management* whereas *quality management and monitoring by information management*

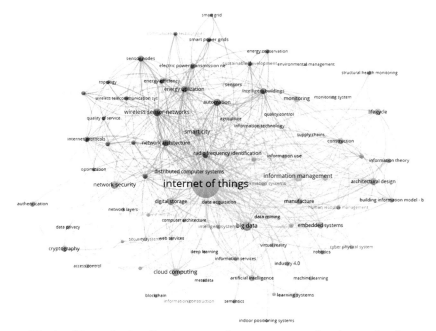

Fig. 4.7: Network visualization map of co-occurrence of author and index keywords focused on IoT for the AECO/FM industry (between 2006 and 2019)

subjects were rare as compared to the former research field. Even though one of the main aims of IoT research for the AECO/FM industry is *lifecycle management by BIM integration,* research on this subject has a low link to IoT-based studies and causes a huge gap in the literature.

4.4 Convergence of BIM and IoT

4.4.1 Convergence Method 1

The three components for the convergence of BIM and IoT are (Lv *et al.,* 2020; Cheng *et al.,* 2020; Heaton *et al.,* 2019; Tang *et al.,* 2019):

1. BIM as an information repository platform serves geometric and spatial data. IoT devices serve real-time data from building consistent occupant patterns, building feedback, social media, weather, financial pricing information, etc. Both contextual BIM model and IoT device-resourced information can be stored in BIM model with IFC format.
2. Continuous sensor-reading records are time-series data that are stored in relational database and Structured Query Language (SQL) is needed for this type of data query (Kazmi *et al.,* 2014).
3. The integration method between the above-mentioned information types.

The following part describes various integration ways of time-series data and contextual information. Relational database and BIM tools' APIs used together is a widely implemented approach (Fig. 4.8). The sequence of activities in the approach is as follows: The real-world time-series data is acquired by sensors to be stored and updated in the relational database. BIM model designed via BIM tools is exported into the relational database format by APIs. Then database schema are defined to clarify the relationship between physical building with sensors and virtual objects as modeled in BIM via GUID or UUID number as a unique identification tool. APIs enable relational database and the BIM model import and export both ways. Custom-built API, third-party processing engine or direct query over SQL database are used to process sensor data queries (Habibi, 2017; Gerrish *et al.*, 2017; Oppermann *et al.*, 2016; Woo *et al.*, 2016; Zhang and Bai, 2015; Arslan *et al.*, 2014; Marzouk and Abdelaty, 2014; Bille *et al.*, 2014; Riaz *et al.*, 2014).

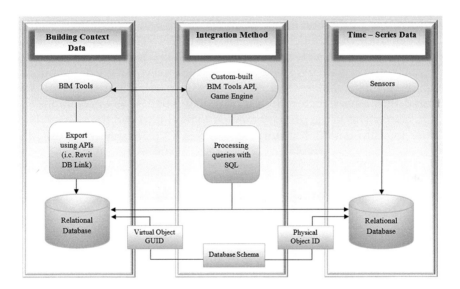

Fig. 4.8: BIM tools' APIs and relational database (adapted from Tang *et al.*, 2019)

There are some advantages and disadvantages in this approach. The advantages are:

1. The existing APIs are able to enable integration. Open database connectivity (ODBC) format can be conformable with external database software. The model data can be exported in ODBC.
2. The model data and sensor data are stored in the relational database which eases the link between these two.

3. With the APIs, it is possible to automatically update the time-series data in BIM tools.
   The disadvantages are:

1. The only parameters that can be exported between projects and families are shared parameters which cause limitations in updating.
2. The sensor data automatically is updated in the relational data. However, files should be manually exported in case of any change in the model.

This approach is not appropriate for complicated BIM models and too many sensors. Because this approach requires physical sensors to be represented by virtual objects manually, which use APIs to export BIM data into the relational database, it does not require intensive IFC programming expertise.

4.4.2 Convergence Method 2

Transforming BIM data into a quarriable database is an effective method for integrating BIM model and sensor data (Fig. 4.9). This approach allows information extraction from various stakeholders' perspectives (Cheng *et al.*, 2020; Kuo *et al.*, 2019; Heaton *et al.*, 2019; Tang *et al.*, 2019). Traditional facility management system stores data in a relational database. The stored BIM data structure into a relational database is effective for the SQL query. This is the primary method to bind BIM and time-series data. After the BIM data is transformed into a SQL quarriable form, the BIM

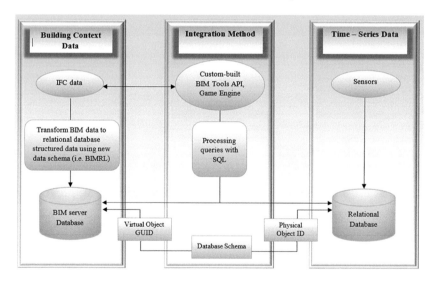

Fig. 4.9: Transformation of BIM data into relational database via new data schema (adapted from Tang *et al.*, 2019)

model becomes a sensor data linkable. The result is virtual sensor objects and physical sensors to be connected via GUID numbers, and sensor-collected data to be mapped as sensor objects' properties (Solihin *et al.*, 2017; Motamedi *et al.*, 2014; Kang and Choi, 2015; Khalili and Chua,.2013).

4.4.3 Convergence Method 3

Creating a new query language is another method for querying sensor data via the IFC or BIM model without using SQL. Processing time-series data queries are developed with newly developed query language (Al-Saeed *et al.*, 2020; Heaton *et al.*, 2019; Alves *et al.*, 2017; Mazairac and Beetz, 2013) (Fig. 4.10).

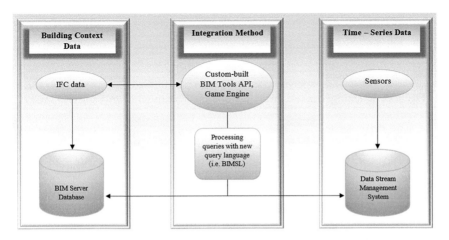

Fig. 4.10: Creating a new query language (adapted from Tang *et al.*, 2019)

4.4.4 Convergence Method 4

AECO/FM activities include the data consisting of geometry, topology, geospatial information, and habitants' behavioral information throughout the building's lifecycle. The heterogeneous data sets' store, share, and use require the integration of Semantic Web technologies and the BIM. Resource Description Format (RDF) is required to represent or tag these datasets. BIM and IoT integration can be realized by linking BIM ontologies, such as ifcOWL to other ontologies, such as semantic sensor networks, smart appliances reference. The contextual data and time-series sensor data should be depicted in a homogeneous format with a semantic web approach for the integration of BIM and IoT (Fig. 4.11). The following process is valid for the semantic web approach in BIM and IoT integration (Quinn *et al.*, 2020; Dibley *et al.*, 2012; Curry *et al.*, 2013; Hu *et al.*, 2016):

a. Contextual information representation into RDF.

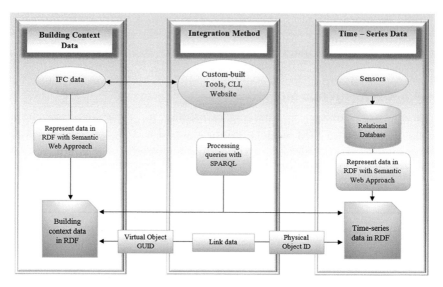

Fig. 4.11: Utilizing semantic web technologies (adapted from Tang *et al.*, 2019)

b. The time-series data received via sensors is acquired from relational database in an RDF format by applying semantic web approach.
c. Unique identification is utilized for linking data silos among different domains.
d. The queries of contextual data or real-time sensor data are actualized by SPARQL, which is used as an RDF query language.
e. The query results are available on applications (GUI, dashboards, command line interface, API, etc.)

4.4.5 Convergence Method 5

In this method, cross-domain data is stored via use of semantic web and relational database together. The implementation process of this method is shown in Fig. 4.12. The process is determined with several steps as listed below:

1. The contextual information of which sensor, building context, and soft building-data representation is in RDF format by semantic web.
2. Retention of time-series data collected by sensors in the relational database.
3. Building context information mapping use of time-series data which is described in RDF with sensor ID referenced.

This approach requires use of two technologies together to get integrated query methods. SPARQL is used to query contextual information in RDF format. Query time-series data is stored into a relational database

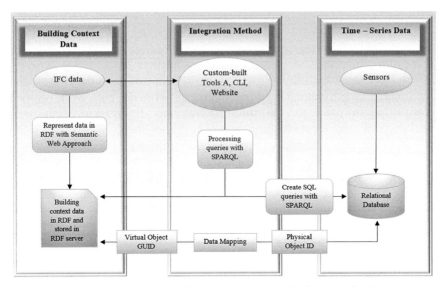

Fig. 4.12: Hybrid approach: Semantic web and relational database
(adapted from Tang *et al.*, 2019)

via SQL. Time-series data and building context information are mapped. Therefore, SPARQL can be used to create SQL queries on RDF data (Hu *et al.*, 2016; McGlinn *et al.*, 2017).

4.5 Digital Twin

4.5.1 The Transformation of a Physical Building into Cyber-Physical System

The convergence of physical buildings and digital models is a big challenge in Cyber-Physical Systems. Smart and intelligent production systems can be realized with Industry 4.0 opportunities to increase efficiency (Schuh *et al.*, 2017; Ribeiro and Bjorkman, 2017; Qin *et al.*, 2016). Grieves coined the idea of Digital Twin in 2003 as a product lifecycle management concept (Grieves, 2014). The Digital Twin is a virtual model of a physical entity. The components of the system are physical building, digital model, and sensors and devices as the connectors. The duplication of physical entities in a digital environment for remote viewing, monitoring, and controlling will affect routine processes and procedures in all organizations. Digital twins, the Internet of Things (IoT), and other cognitive technology integrations may redefine the future vision of the AECO/FM industry as well as the others. Real-time data integration via IoT sensors and devices on the model of the physical system enhances adaptive updating to serve the information for further machine learning and artificial intelligence integration in order to coordinate and automate the physical counterpart according to operational changes. However, the information is needed to be organized, analyzed, and extracted from the collected data by advanced computing mechanisms and algorithms (Hashem *et al.*, 2015; Yi *et al.*, 2015). Unlike the traditional method, designers optimize the design and maximize the performance of the buildings with the help of developed computing industry for real-time control and digitization of the past and present, and forecast the future (Leng and Jiang, 2019; Wang *et al.*, 2016a; Wang *et al.*, 2016b; Xu *et al.*, 2017; Wan *et al.*, 2017; Grieves, 2014). The physical building is mimicked by spontaneously and continuously updated virtual model by multiple sources (Constante, 2018), thereby, sensed data through physical building enhances Digital Twin to estimate and analyze dynamic changes for project lifecycle optimization (Tao *et al.*, 2018).

4.5.2 IoT-integrated Digital Twin

IoT integration on Digital Twin can boost the performance of the physical system. Widespread implementation areas for Digital Twin are performance optimization, safety, user interface, maintenance, construction, and smart cities. Digital Twin implementation has challenges, such as any change in the project design and execution forces direct and immediate changes in the model (Wang and Wang, 2017), design and construction-related information should be acquired, stored, shared, used, and reused throughout the project life cycle via Digital-Twin-driven physical-cyber-social connected production line (Barnaghi *et al.*, 2015; Hussein *et al.*, 2015; Uhlemann *et al.* 2017; Scleich *et al.*, 2017), Big Data analytics for monitor and control changes (Lynch, 2008; Bandaru *et al.*, 2017a; Bandaru *et al.*, 2017b), cross-domain collaboration for technical reasons (Fieldmann and Vogel-Heuser, 2013), security (Kaur *et al.*, 2020).

4.6 Conclusion

This chapter introduces BIM and IoT integration for BIM Digital Twin, which is a platform with the ultimate goal of improvement and optimization of buildings' architectural design, engineering design, construction, operation, and facility management, reducing construction costs and time spent while increasing the overall AECO/FM activities performance. The chapter describes (i) BIM; (ii) IoT; (iii) BIM and IoT-integration methodology under the Digital Twin concept; (iv) Digital Twin, an integrated platform is achieved through an BIM and IoT integration to allow large-scale data, information, and knowledge integration and synchronization, thus allowing better handling and processing. Twinning the virtual information model with real-time data may significantly help in decision making during each phase of the whole building's lifecycle (from initiation to demolition). The integrated BIM Digital Twin platform will enable the interaction of many various stakeholders throughout all phases of a building.

Future works are, however, needed to address various perspectives' framework, further practice of smart DT-enabled process developments, smart DT-enabled buildings/infrastructures' expansion for smart city development, interaction and impact of a DT-enabled building with the diverse stakeholders for spreading a better understanding.

References

Ahankoob, A., Manley, K., Hon, C. and Drogemuller, R. (2018). The impact of building information modeling (BIM) maturity and experience on contractor absorptive capacity. *Architectural Engineering and Design Management*, 14(5). https://doi.org/10.1080/17452007.2018.1467828

Ahmed, A.L. and Kassem, M. (2018). A unified BIM adoption taxonomy: Conceptual development, empirical validation and application. *Automation in Construction*, 96: 103-127. https://doi.org/10.1016/j.autcon.2018.08.017

Allam, Z. and Dhunny, Z.A. (2019). On Big Data, artificial intelligence and smart cities. *Cities*, 89: 80-91. https://doi.org/10.1016/j.cities.2019.01.032

Al-Saeed, Y., Edwards, D.J. and Scaysbrook, S. (2020), Automating construction manufacturing procedures using BIM digital objects (BDOs): Case study of knowledge transfer partnership project in UK. *Construction Innovation*, 20(3): 345-377. https://doi.org/10.1108/CI-12-2019-0141

Alves, M., Carreira, P. and Costa, A.A. (2017). BIMSL: A generic approach to the integration of building information models with real-time sensor data. *Automation in Construction*, 84: 304-314. DOI: 10.1016/j.autcon.2017.09.005

Atlam, H.F. and Wills, G.B. (2020). IoT security, privacy, safety and ethics. *In*: Farsi, M., Daneshkhah, A., Hosseinian-Far, A. and Jahankhani, H. (Eds). Digital Twin Technologies and Smart Cities. Internet of Things (Technology, Communications and Computing), Springer, Cham. https://doi.org/10.1007/978-3-030-18732-3_8

Arslan, M., Riaz, Z., Kiani, A.K. and Azhar, S. (2014). Real-time environmental monitoring, visualization and notification system for construction H&S management. *Electronic Journal of Information Technology in Construction Electron*, 19: 72-91. http://www.itcon.org/2014/4

Azzouz, A. and Hill, P. (2017). Hunting for perfection: How Arup measures BIM maturity on projects worldwide. *Construction Research and Innovation*, 8(2). https://doi.org/10.1080/20450249.2017.1334909

Bandaru, S., Ng, A.H. and Deb, K. (2017). Data mining methods for knowledge discovery in multi-objective optimization: Part A – Survey. *Expert Syst. Appl.*, 70: 139-159. https://doi.org/10.1016/j.eswa.2016.10.015

Bandaru, S., Ng, A.H. and Deb, K. (2017). Data mining methods for knowledge discovery in multi-objective optimization: Part B – New developments and applications. *Expert Syst. Appl.*, 70: 119-138. https://doi.org/10.1016/j. eswa.2016.10.016

Barnaghi, P., Sheth, A., Singh, V. and Hauswirth, M. (2015). Physical-cyber-social computing: Looking back, looking forward. *IEEE Internet Comput.*, 3: 7-11. doi: 10.1109/MIC.2015.65

Bille, R., Smith, S.P., Maund, K. and Brewer, G. (2014). Extending building information models into game engines. Proceedings of the 2014 Conference on Interactive Entertainment – IE2014, ACM, Newcastle, NSW, Australia, pp. 1-8. DOI: 10.1145/2677758.2677764

Burhan, M., Rehman, R.A., Khan, B. and Kim, B.S. (2018). IoT elements, layered architectures and security issues: A comprehensive survey. *Sensors*, 18(2796). DOI: 10.3390/s18092796

Carmona, A.M., Chaparro, A.I., Velasquez, R., Botero-Valencia, J., Castano-Londono, L., Marquez-Viloria, D. and Mesa, A.M. (2019). Instrumentation and data collection methodology to enhance productivity in construction sites using embedded systems and IoT technologies. *In*: Mutis, I. and Hartmann, T. (Eds). Advances in Informatics and Computing in Civil and Construction Engineering. Springer, Cham. https://doi.org/10.1007/978-3-030-00220-6_76

Chang, K.M., Dzeng, R.J. and Wu, Y.J. (2018). An automated IoT visualization BIM platform for decision support in facilities management. *Appl. Sci.*, 8(7): 1086. https://doi.org/10.3390/app8071086

Chen, C. and Tang, L. (2019). BIM-based integrated management workflow design for schedule and cost planning of building fabric maintenance. *Automation in Construction*, 107(November): 102944. https://doi.org/10.1016/j.autcon. 2019. 102944

Cheng, J.C.P., Chen, W., Chen, K. and Wang, Q. (2020). Data-driven predictive maintenance planning framework for MEP components based on BIM and IoT using machine learning algorithms. *Automation in Construction*, 112: 103087. https://doi.org/10.1016/j.autcon.2020.103087

Constante, T.A.D.S.L. (2018). *Contribution for a Simulation Framework for Designing and Evaluating Manufacturing Systems*. MSc. Thesis, Faculty of Engineering, University of Porto.

Costin, A., Wehle, A. and Adibfar, A. (2019). Leading indicators – A conceptual IoT-based framework to produce active leading indicators for construction safety. *Safety*, 5(4): 86. https://doi.org/10.3390/safety5040086

Curry, E., O'Donnell, J., Corry, E., Hasan, S., Keane, M. and O'Riain, S. (2013). Linking building data in the cloud: Integrating cross-domain building data using linked data. *Advance Engineering Informatics*, 27(2): 206-219. DOI: 10.1016/j.aei.2012.10.003

Dibley, M., Li, H., Rezgui, Y. and Miles, J. (2012). An ontology framework for intelligent sensor-based building monitoring. *Automation in Construction*, 28: 1-14, DOI: 10.1016/j.autcon.2012.05.018

Edirisinghe, R. and Woo, J. (2020). BIM-based performance monitoring for smart building management. *Facilities*, vol. ahead-of-print. https://doi. org/10.1108/F-11-2019-0120

Feldmann, S. and Vogel-Heuser, B. (2013). Änderungsszenarien in der Automatisierungstechnik-Herausforderungen und interdisziplinäre Auswirkungen. *Engineering von der Anforderung bis zum Betrieb*, 3: 95.

Gamil, Y., Abdullah, M.A., Rahman, I.A. and Asad, M.M. (2020). Internet of Things in construction industry revolution 4.0: Recent trends and challenges in the Malaysian context. *Journal of Engineering, Design and Technology*, ahead of print. https://doi.org/10.1108/JEDT-06-2019-0164

Gerrish, T., Ruikar, K., Cook, M., Johnson, M., Phillip, M. and Lowry, C. (2017). BIM application to building energy performance visualisation and management: Challenges and potential. *Energy and Buildings*, 144: 218-228. DOI: 10.1016/j. enbuild.2017.03.032

Grieves, M. (2014). *Digital Twin: Manufacturing Excellence through Virtual Factory Replication*, white paper, 1: 1-7.

Habibi, S. (2017). Micro-climatization and real-time digitalization effects on energy efficiency based on user behavior. *Building and Environment*, 114: 410-428. DOI: 10.1016/j.buildenv.2016.12.039

Häikiö, J., Kallio, J., Mäkelä, S.M. and Keränen, J. (2020). IoT-based safety monitoring from the perspective of construction site workers. *International Journal of Occupational and Environmental Safety*, 4(1): 1-14.

Hashem, I.A.T., Yaqoob, I., Anuar, N.B., Mokhtar, S., Gani, A. and Khan, S.U. (2015). The rise of big data on cloud computing: Review and open research issues. *Inf. Syst.*, 47: 98-115. https://doi.org/10.1016/j.is.2014.07.006

Hussein, D., Park, S., Han, S.N. and Crespi, N. (2015). Dynamic social structure of things: A contextual approach in CPSS. *IEEE Internet Comput.*, 19(3): 12-20. DOI: 10.1109/MIC.2015.27

He, W., Li, S. and Xu, L.D. (2014). Internet of Things in industries: A survey. *IEEE Transactions on Industrial Informatics*, 10(4): 2233-2243. DOI: 10.1109/TII.2014.2300753

Heaton, J., Parlikad, A.K. and Schooling, J. (2019). Design and development of BIM models to support operations and maintenance. *Computers in Industry*, 111: 172-186. https://doi.org/10.1016/j.compind.2019.08.001

Hu, S., Corry, E., Curry, E., Turner, W.J.N. and O'Donnell, J. (2016). Building performance optimisation: A hybrid architecture for the integration of contextual information and time-series data. *Automation in Construction*, 70: 51-61. DOI: 10.1016/j.autcon.2016.05.018

Hwang, B.G., Zhao, X. and Yang, K.W. (2019). Effect of BIM on rework in construction projects in Singapore: Status quo, magnitude, impact, and strategies. *Journal of Construction Engineering and Management*, 145(2). https://doi.org/10.1061/(ASCE)CO.1943-7862.0001600

Joel, M.R., Ebenezer, V., Karthik, N. and Rajkumar, K. (2019). Advance dynamic network system of Internet of Things. *International Journal of Recent Technology and Engineering (IJRTE)*, 8(3). DOI: 10.35940/ijrte.C5657.098319

Juan, Y.K., Lai, W.Y. and Shih, S.G. (2017). Building information modeling acceptance and readiness assessment in Taiwanese architectural firms. *Journal of Civil Engineering and Mangement*, 23(3). https://doi.org/10.3846/13923730.2015.1128480

Kang, T.W. and Choi, H.S. (2015). BIM perspective definition metadata for interworking facility management data. *Advance Engineering Informatics*, 29: 958-970. DOI: 10.1016/j.aei.2015.09.004

Kassem, M. and Succar, B. (2017). Macro BIM adoption: Comparative market analysis. *Automation in Construction*, 81: 286-299. https://doi.org/10.1016/j.autcon.2017.04.005

Kaur, M.J., Mishra, V.P. and Maheshwari, P. (2020) The convergence of Digital Twin, IoT, and machine learning: Transforming data into action. *In*: Farsi, M., Daneshkhah, A., Hosseinian-Far, A. and Jahankhani, H. (Eds). Digital Twin Technologies and Smart Cities. Internet of Things (Technology, Communications and Computing). Springer, Cham. https://doi.org/10.1007/978-3-030-18732-3_1

Kazmi, H., O'grady, M.J., Delaney, D.T., Ruzzelli, A.G. and O'hare, G.M.P. (2014). A review of wireless-sensor-network-enabled building energy management systems. *ACM Trans. Sens. Networks (TOSN)*, 10(66): 1–66. 43https://doi.org/10.1145/2532644

Khalili, A. and Chua, D.K.H. (2013). IFC-based graph data model for topological queries on building elements. *Journal of Computing in Civil Engineering*, 29(3). Article: 04014046, DOI: 10.1061/(ASCE)CP.1943-5487.0000331

Kong, L. and Ma, B. (2020). Intelligent manufacturing model of construction industry based on Internet of Things technology. *Int. J. Adv. Manuf. Technol.*, 107: 1025-1037. https://doi.org/10.1007/s00170-019-04369-8

Kortuem, G., Kawsar, F., Sundramoorthy, V. and Fitton, D. (2010). Smart objects as building blocks for the Internet of Things. *IEEE Internet Computing*, 14(1): 44-51 DOI: 10.1109/MIC.2009.143

Košťál, K., Helebrandt, P., Belluš, M., Ries, M. and Kotuliak, I. (2019). Management and monitoring of IoT devices using blockchain. *Sensors*, 19(4): 856. https://doi.org/10.3390/s19040856

Kuo, W.L., Lee, H.X. and Hsieh, S.H. (2019). Designing a database scheme for supporting visual management of variable parameters in BIM models. ASCE International Conference on Computing in Civil Engineering 2019, June 17-19, Atlanta, Georgia. https://doi.org/10.1061/9780784482421.054

Lai, H., Deng, X. and Chang, T.Y.P. (2019). BIM-based platform for collaborative building design and project management. *Journal of Computing in Civil Engineering*, 33(3). DOI: 10.1061/(ASCE)CP.1943-5487.0000830

Lee, I. and Lee, K. (2015). The Internet of Things (IoT): Applications, investments, and challenges for enterprises. *Business Horizons*, 58(4): 431-440. DOI: 10.1016/j.bushor.2015.03.008

Leng, J. and Jiang, P. (2019). Dynamic scheduling in RFID-driven discrete manufacturing system by using multi-layer network metrics as heuristic information. *J. Intell. Manuf.*, 30: 979-994.

Liang, C., Lu, W., Rowlinson, S. and Zhang, X. (2016). Development of a multifunctional BIM maturity model. *Journal of Construction Engineering and Management*, 142(11). https://doi.org/10.1061/(ASCE)CO.1943-7862.0001186

Lu, W., Chen, K., Zetkulic, A. and Liang, C. (2018). Measuring building information modeling maturity: A Hong Kong case study. *International Journal of Construction Management*. https://doi.org/10.1080/15623599.2018.1532385

Lv, Z., Li, X., Lv, H. and Xiu, W. (2020). BIM Big Data Storage in WebVRGIS. *In*: IEEE Transactions on Industrial Informatics, 16(4): 2566-2573. April. DOI: 10.1109/TII.2019.2916689

Lynch, C. (2008). Big Data: How do your data grow? *Nature*, 455(7209): 28. https://doi.org/10.1038/455028a

Mahamadu, A.M., Manu, P., Mahdjoubi, L., Booth, C., Aigbavboa, C. and Abanda, F.H. (2019). The importance of BIM capability assessment: An evaluation of post-selection performance of organisations on construction projects. *Engineering, Construction and Architectural Management*, 27(1): 24-48. https://doi.org/10.1108/ECAM-09-2018-0357

Mazairac, W. and Beetz, J. (2013). BIMQL – An open query language for building information models. *Advance Engineering Informatics*, 27(4): 444-456. DOI: 10.1016/j.aei.2013.06.001

Marzouk, M. and Abdelaty, A. (2014). Monitoring thermal comfort in subways using building information modeling. *Energy and Buildings*, 84: 252-257. DOI: 10.1016/j.enbuild.2014.08.006

McGlinn, K., Yuce, B., Wicaksono, H., Howell, S. and Rezgui, Y. (2017). Usability evaluation of a web-based tool for supporting holistic building energy management. *Automation in Construction*, 84: 154-165. DOI: 10.1016/j.autcon.2017.08.033

Motamedi, A., Hammad, A. and Asen, Y. (2014). Knowledge-assisted BIM-based visual analytics for failure root cause detection in facilities management. *Automation in Construction*, 43: 73-83. DOI: 10.1016/j.autcon.2014.03.012

Nagashree, R.N., Rao, V. and Aswini, N. (2014). Near field communication. *Int. J. Wireless and Microwave Technologies (IJWMT)*, 4: 20-30. DOI: 10.5815/ijwmt.2014.02.03

Niu, Y., Anumba, C. and Lu, W. (2019). Taxonomy and deployment framework for emerging pervasive technologies in construction projects. *Journal of Construction Engineering and Management*, 145(5). https://doi.org/10.1061/(ASCE)CO.1943-7862.0001653

Noor, M.B.M. and Hassan, W.H. (2019). Current research on Internet of Things (IoT) security: A survey. *Computer Networks*, 148: 283-294. https://doi.org/10.1016/j.comnet.2018.11.025

Opentechdiary (2015). *Internet of Things World Europe*. Retrieved from: https://opentechdiary.wordpress.com/2015/07/16/a-walk-through-internet-of-things-iot-basics- part-2/

Oppermann, L., Shekow, M. and Bicer, D. (2016). Mobile cross-media visualizations made from building information modeling data. Proceedings of the 18th International Conference on Human-Computer Interaction with Mobile Devices and Services Adjunct – Mobile HCI '16, ACM, Florence, Italy. 823-830, DOI: 10.1145/2957265.2961852

Ozturk, G.B. (2020). Interoperability in building information modeling for AECO/FM industry. *Automation in Construction*, 113. Art. No. 103122. DOI: 10.1016/j.autcon.2020.103122

Papadonikolaki, E., Oel, C.V. and Kagioglou, M. (2019). Organizing and managing boundaries: A structural view of collaboration with Building Information Modeling (BIM). *International Journal of Project Management*, 37(3): 378-394. https://doi.org/10.1016/j.ijproman.2019.01.010.

Peng, C. and Huang, J. (2016). A home energy monitoring and control system based on ZigBee technology. *International Journal of Green Energy*, 13(15): 1615-1623. DOI: 10.1080/15435075.2016.1188102

Puyan, A.Z., Wei, L., Dee, A., Pottinger, R. and Staub-French, S. (2019). BIM-CITYGML data integration for modern urban challenges. *Journal of Information Technology in Construction (ITcon)*, 24: 318-340. http://www.itcon.org/2019/17

Qin, J., Liu, Y. and Grosvenor, R. (2016). A categorical framework of manufacturing for Industry 4.0 and beyond. *Procedia Cirp.*, 52: 173-178. https://doi.org/10.1016/j.procir.2016.08.005

Quinn, C., Shabestari, A.Z., Misic, T., Gilani, S., Litoiu, S. and McArthur, J.J. (2020). Building automation system – BIM integration using a linked data structure. *Automation in Construction*, 118: 103257. https://doi.org/10.1016/j.autcon.2020.103257

Raji, R.S. (1994). Smart networks for control. *IEEE Spectrum*, 31(6): 49-55. DOI: 10.1109/6.284793

Riaz, Z., Arslan, M., Kiani, A.K. and Azhar, S. (2014). CoSMoS: A BIM and wireless sensor based integrated solution for worker safety in confined spaces. *Automation in Construction*, 45: 96-106. DOI: 10.1016/j.autcon.2014.05.010

Ribeiro, L. and Björkman, M. (2018). Transitioning from standard automation solutions to cyber-physical production systems: An assessment of critical conceptual and technical challenges. *IEEE Syst. J.*, 12(4): 1-13. doi: 10.1109/JSYST.2017.2771139

Rogage, K., Clear, A., Alwan, Z., Lawrence, T. and Kelly, G. (2019). Assessing building performance in residential buildings using BIM and sensor data. *International Journal of Building Pathology and Adaptation*, 38(1): 176-191. https://doi.org/10.1108/IJBPA-01-2019-0012

Salleh, A., Aziz, A., Abidin, M.Z., Misran, M.H. and Mohamad, N.R. (2013). Development of greenhouse monitoring using wireless sensor network through ZigBee technology. *International Journal of Engineering Science Invention*, ISSN 2(7): 6-12.

Sanchez, A.X. and Joske, W. (2016). Benefits dictionary. pp. 103-204. *In*: A.X. Sanchez, K.D. Hampson, and S. Vaux (eds.). Delivering Value with BIM: A Whole-of-Life Approach. Routledge, Oxon, UK.

Schleich, B., Anwer, N., Mathieu, L. and Wartzack, S. (2017). Shaping the Digital Twin for design and production engineering. *CIRP Ann.*, 66(1): 141-144. https://doi.org/10.1016/j.cirp.2017.04.040

Schuh, G., Anderl, R., Gausemeier, J., Hompel, M.T. and Wahlster W. (2017). Industrie 4.0 maturity index. *Managing the Digital Transformation of Companies*, Munich: Herbert Utz.

Siebelink, S., Voodijk, J.T. and Adriaanse, A. (2018). Developing and testing a tool to evaluate BIM maturity: Sectoral analysis in the Dutch Construction Industry. *Journal of Construction Engineering and Management*, 144(8). https://doi.org/10.1061/(ASCE)CO.1943-7862.0001527

Shou, W., Wang, J., Wang, X. and Chong, H.Y. (2015). A comparative review of building information modeling implementation in building and infrastructure industries. *Arch. Comput. Methods Eng.*, 22(2): 291-308.

Smits, W., Buiten, M.V. and Hartmann, T. (2016). Yield-to-BIM: Impacts of BIM maturity on project performance. *Building Research & Information*, 45(3). https://doi.org/10.1080/09613218.2016.1190579

Solibri (2016). *About BIM and IFC*. Retrieved 30 June, 2020, from tp://www.solibri.com/support/bim-ifc/

Solihin, W., Eastman, C., Lee, Y.C. and Yang, D.H. (2017). A simplified relational database schema for transformation of BIM data into a query-efficient and spatially-enabled database. *Automation in Construction*, 84: 367-383. DOI: 10.1016/j.autcon.2017.10.002

Succar, B., Sher, W. and Williams, A. (2013). An integrated approach to BIM competency assessment, acquisition and application. *Automation in Construction*, 35: 174-189.

Sung, W.T., Hsiao, S.J. and Shih, J.A. (2019). Construction of indoor thermal comfort environmental monitoring system based on the IoT architecture. *Journal of Sensors*. https://doi.org/10.1155/2019/2639787

Talaviya, G., Ramteke, R. and Shete, A.K. (2013). Wireless fingerprint-based college attendance system using Zigbee technology. *International Journal of Security, Privacy and Trust Management (IJSPTM)*, 5(4): 11-17.

Tang, S., Shelden, D.R., Eastman, C.M., Bozorgi, P.P. and Gao X. (2019). A review of building information modeling (BIM) and the internet of things (IoT) devices integration: Present status and future trends. *Automation in Construction*, 101: 127-139. DOI: 10.1016/j.autcon.2019.01.020

Tao, F., Cheng, J., Qi, Q., Zhang, M., Zhang, H. and Sui, F. (2018). Digital Twin-driven product design, manufacturing and service with Big Data. *The International Journal of Advanced Manufacturing Technology*, 94(4): 3563-3576. DOI: 10.1007/s00170-017-0233-1

Tsilimantou, E., Delegou, E.T., Nikitakos, I.A., Ioannidis, C. and Moropoulou, A. (2020). GIS and BIM as integrated digital environments for modeling and monitoring of historic buildings. *Appl. Sci.*, 10(3): 1078. https://doi.org/10.3390/app10031078

Uhlemann, T.H.J., Lehmann, C. and Steinhilper, R. (2017). The Digital Twin: Realizing the cyber-physical production system for Industry 4.0. *Procedia Cirp.*, 61: 335-340. https://doi.org/10.1016/j.procir.2016.11.152

Wan, J., Tang, S., Li, D., Wang, S., Liu, C., Abbas, H. and Vasilakos, A.V. (2017). A manufacturing Big Data solution for active preventive maintenance. *IEEE Trans. Industr. Inf.*, 13(4): 2039-2047. doi: 10.1109/TII.2017.2670505

Wang, W., He, G. and Wan, J. (2011). Research on Zigbee wireless communication technology. *In*: Proceedings of the 2011 International Conference on Electrical and Control Engineering (ICECE), Yichang, China, 16-18 September 2011, 1245-1249. DOI: 10.1109/ICECENG.2011.6057961

Wang, S., Wan, J., Li, D. and Zhang, C. (2016b). Implementing smart factory of Industry 4.0: An outlook. *Int. J. Distrib. Sens. Netw.* 12(1): 3159805. https://doi.org/10.1155/2016/3159805

Wang, S., Wan, J., Zhang, D., Li, D. and Zhang, C. (2016). Towards smart factory for Industry 4.0: A self-organized multi-agent system with Big Data-based feedback and coordination. *Comput. Netw.*, 101: 158-168. https://doi.org/10.1016/j.comnet.2015.12.017

Wang, X.V. and Wang, L. (2017). A cloud-based production system for information and service integration: An Internet of Things case study on waste electronics. *Enterp. Inf. Syst.*, 11(7): 952-968. https://doi.org/10.1080/17517575.2016.1215539

Wigmore, I. (2014). Internet of Things (IoT). *Tech Target*. Retrieved from https://internetofthingsagenda.techtarget.com/definition/Internet-of-Things-IoT

Whitmore, A., Agarwal, A. and Da Xu, L. (2015). The Internet of Things – A survey of topics and trends. *Information Systems Frontiers*, 17: 261-274. doi 10.1007/s10796-014-9489-2

Woo, J.H., Peterson, M.A. and Gleason, B. (2016). Developing a virtual campus model in an interactive game-engine environment for building energy benchmarking. *Journal of Computing in Civil Engineering*, 30(5): Article C4016005, DOI: 10.1061/(ASCE)CP.1943-5487.0000600

Wu, I.C. and Liu, C.C. (2020). A visual and persuasive energy conservation system based on BIM and IoT technology. *Sensors*, 20(1): 139. https://doi.org/10.3390/s20010139

Wu, C., Xu, B., Mao, C. and Li, X. (2017). Overview of BIM maturity measurement tools. *Journal of Information Technology in Construction (ITcon)*, 22: 34-62. http://www.itcon.org/2017/3

Xu, Y., Sun, Y., Wan, J., Liu, X. and Song, Z. (2017). Industrial Big Data for fault diagnosis: Taxonomy, review, and applications. *IEEE Access*, 5: 17368-17380. DOI: 10.1109/ACCESS.2017.2731945

Yi, S., Li, C. and Li, Q. (2015). A survey of fog computing: Concepts, applications and issues. *In*: Proceedings of the 2015 workshop on Mobile Big Data, ACM, pp. 37-42. https://doi.org/10.1145/2757384.2757397

Yilmaz, G., Akcamete, A. and Demirors, O. (2019). A reference model for BIM capability assessments. *Automation in Construction*, 101: 245-263. https://doi.org/10.1016/j.autcon.2018.10.022

Zhai, Y., Chen, K., Zhou, J.X., Cao, J., Lyu, Z., Jin, X., Shen, G.Q.P., Lu, W. and Huang, G.Q. (2019). An Internet of Things-enabled BIM platform for modular integrated construction: A case study in Hong Kong. *Advanced Engineering Informatics*, 42: 100997. https://doi.org/10.1016/j.aei.2019.100997

Zhao, L., Liu, Z. and Mbachu, J. (2019). Development of intelligent prefabs using IoT technology to improve the performance of prefabricated construction projects. *Sensors*, 19(19): 4131. https://doi.org/10.3390/s19194131

Zhang, Y. and Bai, L. (2015). Rapid structural condition assessment using radio frequency identification (RFID)-based wireless strain sensor. *Automation in Construction*, 54: 1-11. doi: 10.1016/j.autcon.2015.02.013

Zhang, D., Yang, L.T., Chen, M., Zhao, S., Guo, M. and Zhang, Y. (2016). Real-time locating systems using active RFID for Internet of Things. *IEEE Systems Journals*, 10(3): 1226-1235. DOI: 10.1109/JSYST.2014.2346625

Zhou, Z. (2012). Application of Internet of Things in agriculture products supply chain management. *In*: Proceedings of the 2012 International Conference on Control Engineering and Communication Technology (ICCECT), Liaoning, China, 7-9 December 2012, pp. 259-261.

Zillner, T. (2018). *Zigbee Exploited – The Good, the Bad and the Ugly*. Available online: https://www.blackhat.com/ docs/us-15/materials/us-15-Zillner-ZigBee-Exploited-The-Good-The-Bad-And-The-Ugly.pdf (accessed on 6 January 2018).

The Integration of Building Information Modeling (BIM) and Immersive Technologies (ImTech) for Digital Twin Implementation in the AECO/FM Industry

Gözde Basak Özturk

Faculty of Engineering, Civil Engineering Department, Aydin Adnan Menderes University, Central Campus P. Box: 09010 Aydın, Turkey

5.1 Introduction

Over the years, the AECO/FM industry has struggled with data acquisition, store, share, use, and reuse of enormous amounts of project-related information, which causes poor performance (Dainty *et al.*, 2006). It is crucial to managing project and facility information accurately and timely among project stakeholders for effective and efficient processes. Unless robust information management occurs throughout the project lifecycle, productivity decreases with causing time and cost overruns, waste increases, quality decreases, monitoring and controlling sufficiency decreases, errors increases, and communication lacks.

Recently, Information and Communication Technology (ICT) has gained much attention as a key driver of change in the architecture, engineering, construction, operation, and facility management (AECO/FM) industry. However, the adoption of ICT is relatively slow in the AECO/FM industry when compared to other industries (McKinsey Global Institute, 2016). The reason for the slow adoption rate is lack of

Email: gbozturk@adu.edu.tr

technical capability, skilled professionals, and resistance to systematical and methodological change. Developments in ICT have provided numerous opportunities for the AECO/FM industry. Building information modeling (BIM) is one of these technologies, which serves as a central data repository that can be utilized for data acquisition, store, share, use, and reuse of information about a facility. It is currently regarded as an essential tool in managing the lifecycle of a construction project right from initiation to its demolition. BIM can be determined as a technology, as well as a process. BIM is considered to be a multi-dimensional digital representation of the physical and functional characteristics of a project and provides the opportunity for active information exploitation and exploration among project stakeholders. Various types of information can be added on the BIM platform, in which different dimensions, such as the fourth dimension (time information), fifth dimension (cost information), sixth dimension (sustainability information), seventh dimension (facility management information) are set with each new type of information. Various researchers have already proposed implementation of BIM concepts into many subjects, such as time, cost, scope, quality (Mirshokraei *et al.*, 2019), communications, human resource, risk, integration, procurement, stakeholder management, collaboration, efficiency, productivity, design, construction, operation, maintenance and so on. In this sense, many researchers have proposed the combination of BIM with other technologies to expand the potential benefit of this technology. There are many studies, involving different techniques and technologies, such as cognitive technologies, Internet of Things (IoT), machine learning, artificial intelligence (AI), immersive technologies virtual reality, augmented reality, mixed reality, extended reality), laser-scan point clouds (Akinci *et al.*, 2006; Tang *et al.*, 2009; Tang *et al.*, 2011; Anil *et al.*, 2013; Bosche *et al.*, 2015), indoor positioning through magnetic fields and wi-fi signals, personal digital assistants (Kimoto *et al.*, 2005), radio frequency identification (Jasekskis *et al.*, 1995; Wang, 2008; Moon and Yang, 2010), and mobile devices (Davies and Harty, 2013; Ma *et al.*, 2018).

In the last decade, Immersive Technologies (ImTech) have drawn considerably broad attention in the AECO/FM community (Wang *et al.*, 2018). There are many advantages and application fields of ImTech (Wang *et al.*, 2012; Golparvar-Fard *et al.*, 2011; Wang *et al.*, 2013; Wang *et al.*. 2014; Park *et al.* 2013; Kwon *et al.*, 2014). ImTech allow overlaying real-world objects and virtual objects to present needed information. ImTech implementation areas in the AECO/FM industry are visualization, simulation, information modeling, progress monitoring, communication, collaboration, progress monitoring, education, training, information access, evaluation, safety, and inspection (Rankohi and Waugh, 2013). ImTech and BIM are complementary technologies (Wang *et al.*, 2012). The convergence of BIM and ImTech can maximize the potentials of BIM by increasing automation of processes, improve the decision-making process, and provide real-time access to information. The use of BIM and ImTech together can be summarized as follows:

Time management, project control, procurement, monitoring, design control, visualization of design during construction, construction defect management (Park *et al.*, 2013), maintenance (Alam *et al.*, 2017; Neges *et al.*, 2017; Koch *et al.*, 2014; Ong and Zhu, 2013; Lee and Akin, 2011; Liu and Seipel, 2018, and Fiorentiona *et al.*, 2014), efficiency and effectiveness improvement (Chu *et al.*, 2018), facility management (Irizarry *et al.*, 2013). Various researches concluded BIM and ImTech integration proposes a great potential by facilitating 3D building data representation through simulation onto real-world data (Meza *et al.*, 2014; Wang *et al.*, 2014; Hou *et al.*, 2015; Wang *et al.*, 2013; Ammari and Hammad, 2014).

5.2 Building Information Modeling (BIM) and Immersive Technologies (ImTech)

5.2.1 The Potential of ImTech

Many industries are affected by the emergence of immersive technologies, which transformed human beings' interaction with visual information. Immersive technologies are determined as one of the most important technologies by Gartner strategic technology trends in 2019 (Panneta, 2018). A most important focus in BIM research will be immersive technologies as the future research trend suggested by Ozturk (2020). Industries that benefit from immersive technologies are game, entertainment, aerospace, logistics, retail, marketing, tourism, education, sports, training, and such (Research and Markets, 2018). Immersive technology market share is expected to be $80 billion by 2025 (Goldman Sachs, 2016). A recent investigation shows that the market share of immersive technologies is expected to be $94 billion by 2023 (Research and Market, 2018). Kaiser and Scatsky (2017) state that companies investing in immersive technologies incredibly increased (230 per cent) their venture capital investment between 2016 and 2017.

5.2.2 BIM as a Platform to Integrate ImTech

Despite its high potential, immersive technology adoption in AECO/FM activities is low because of the nature of the industry. Manyika *et al.* (2015) compared 22 industries in terms of digitalization and the results of the study revealed that the construction industry is the slowest among all. The low adoption rate is caused by many complex and interrelated factors.

BIM has been on the agenda of the AECO/FM industry almost for the last two decades. The BIM technology allows the digitalization of buildings' models with integrated nD information on it to serve for the building's lifecycle. BIM has the capability of object-oriented information structuring and exchange, which allows all stakeholders to access and integrate various information related to various professions with centralized information on the model (Irizarry *et al.*, 2013). BIM, as a digital technology platform, gives an opportunity to integrate technologies, such as immersive technologies, IoT, machine learning, artificial intelligence, and others. Full collaboration among stakeholders is elaborated via cloud-based BIM systems (Matthews *et al.*, 2015), which allow accessing of the centralized database through mobile devices. However, cloud-based systems still cannot stop the tendency in archiving and have visualization problems (Li *et al.*, 2018). Immersive technologies integration is needed to extract acquired building data for increasing BIM effectiveness (Meza *et al.*, 2014). Thus immersive technologies can accelerate decision making by enabling the user interaction with contextual data that is augmented virtual information in real-time (Dunston and Wang, 2003; Hou *et al.*, 2015).

There are many advantages of BIM implementation in the AECO/FM industry, in which visualization occupies an important share. Several researches had been carried out (Wu and Hseih, 2012; Hu and Guo, 2011; Mhalas *et al.*, 2013) focusing on visualization-integrated BIM solutions aiming, integration management, education, energy management, etc. Research related to visualization integration on BIM examines a wide range of project lifecycle problems occurring throughout the design, construction, operation, and facility management activities. The design process researches consist of game engine use in design visualization, design education, interoperability, building-to-city integration, etc. (Yan *et al.*, 2011; Chen *et al.*, 2013; Hagedorn and Döllner, 2007). The construction process researches investigate supply-chain management, monitoring, worker-oriented visualization, change management, information management, information share, safety, site management, automated construction progress monitoring, automated safety rules checking, lean process management, and green approach (Han *et al.*, 2013, Konig *et al.*, 2012; Merivirta *et al.*, 2011; Siu *et al.*, 2013; Kim and Kim 2012; Zhang *et al.*, 2013; Sachs *et al.*, 2013; Memarzadeh and Golparvar-Fard, 2012). The operation and facility-management researches analyze facility management for complex buildings, GIS-based visualization, etc. (Wang *et al.*, 2013; Liu and Issa, 2012).

Immersive technologies are studied in many various fields and industries because of their usefulness in visualization in terms of overlaying real-world information and computer graphics for several purposes (Kim and Irizarry, 2020; Nnaji and Karakhan, 2020; Muhammad *et al.*, 2019; Klatzky *et al.*, 2008; Lamounier *et al.*, 2010; Shekhar *et al.*, 2010; Simpfendörfer *et al.* 2011; Zhu *et al.*, 2004; Pereira *et al.*, 2011; Wakeman *et al.*, 2011; Brown and Barros, 2013; Billinghurst and Dünser, 2012; Hsiao *et al.*, 2012; Matsutomo *et al.*, 2012; Kamarainen *et al.*, 2013; Huynh *et al.*, 2009; Burke *et al.*, 2010; Botella *et al.*, 2011). The immersive technology research trend is accelerating due to the various industries' widely growing interest in the subject. The AECO/FM industry is also one which is intending to facilitate the technology in understanding building research from design to the facility management process. Immersive technologies are investigated by many researchers with a focus on GPS integration, education, monitoring, site management, nD information integration, productivity, risk management, and sustainability (Kim and Irizarry, 2020; Nnaji and Karakhan 2020; Muhammad *et al.*, 2019, Behzadan and Kamat, 2009; Bae *et al.*, 2013; Woodward and Hakkarainen, 2011; Hou and Wang *et al.*, 2013; Hou *et al.*, 2013; Su *et al.*, 2013; Sheng *et al.*, 2011).

5.3 Immersive Technologies (ImTech)

Immersive technologies, such as virtual reality (VR), augmented reality (AR), mixed reality (MR), and extended reality (XR) are visualization technologies that change the way of human's interaction with visual information. Immersive technologies become more extensively used in many industries with the rapid adoption of improved technologies (Fig. 5.1; Table 5.1). The main application areas for immersive technologies are in the gaming and entertainment industries. However, sectoral activities, such as health care, construction, marketing, education, tourism, training, and sports have benefited from immersive technologies (Getuli *et al.*, 2020; Abbas *et al.*, 2019; Research and Markets, 2018).

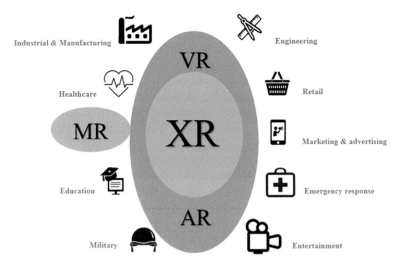

Fig. 5.1: Immersive technology use in various industries

Table 5.1: Main Research Subjects about Immersive Technology Research in the AECO/FM Industry

Research subject	References
Monitoring	Rahimian *et al.*, 2020
Lean	Alizadehsalehi, 2019b
Education	Alizadehsalehi, 2019a; Portman *et al.*, 2015; Sampaio and Martins, 2014
Safety	Li *et al.*, 2018; Shi *et al.*, 2018; Olorunfemi *et al.*, 2018; Bosch-Sijtsema *et al.*, 2019; Azhar, 2017; Klempous *et al.*, 2017; Froehlich and Azhar, 2016; Zhao and Lucas, 2015
Collaborative Decision Making	Du *et al.*, 2018; Du *et al.*, 2017
Training	Mo *et al.*, 2018; Froehlich and Azhar, 2016; Sachs *et al.*, 2013
Prefabrication	Chalhoub and Ayer, 2018
Site Survey	Chalhoub *et al.*, 2018
Effectiveness Evaluation	Chu *et al.*, 2018
Operation	Bosch-Sijtsema *et al.*, 2019
Maintenance	Bosch-Sijtsema *et al.*, 2019
Productivity	Bosch-Sijtsema *et al.*, 2019
Design Development and Improvement	Paes *et al.*, 2017; Haggard, 2017
Building Energy	Niu *et al.*, 2016
Environmental Representation	Kuliga *et al.*, 2015
Communication	Wang *et al.*, 2014

Since immersive technologies have high capability in simulation and visualization, they are used for various purposes in the AECO/FM industry. The emergence of BIM technology enriches the nD modeling capability and information integration. Along with this development, VR solutions ease communication and visualization over the building prototype in the virtual environment. Developments in immersive technology use in the AECO/FM industry have a wide range of research areas and are in need of further attention. The integration of BIM and immersive technologies has achieved considerable success. There are researches focusing on automation in construction monitoring (Rahimian *et al.*, 2020; Kopsida and Brilakis, 2020; Love *et al.*, 2011) and design validation for complex projects (Jin *et al.*, 2020; Wolfartsberger, 2019; Wang *et al.*, 2014). However, construction planning and management field research shows lack of application as seen in the literature. There are four types of immersive technology further explained in the following pages.

5.3.1 Virtual Reality

Virtual reality is a computer technology that simulates reality by replicating the real-world or creating a newly designed one by generating realistic sensational effects by adding sound, visions, and others. The VR environment enhances the user to perceive via all the five senses to mimic an entirely virtual world. Virtual reality technology is mostly used in the game industry. However, developments in the technology and interoperability allow the technology to be used in other industries, such as the AECO/FM industry (Alizadehsalehi *et al.*, 2019a; Moon *et al.*, 2019). The further progress in research on technology is promising and is expected to be realized in the near future.

5.3.2 Augmented Reality

Sound, graphics, GPS data, and video as computer-generated sensory inputs are used to supplement real-world environment elements for augmented reality which is a direct or indirect view. AR provides flexibility in what is wanted to be designed on top of the real-world; thereby, the existing reality is enriched with the designed one by utilizing devices, such as mobiles, tablets, and custom headsets through a camera and applications that are used for overlaying digital content on the physical world data (Kivrak and Arslan, 2019; Cabero-Almenara and Roig-Vila, 2019).

Augmented reality and building information modeling are new technologies in construction. Augmented reality is a field of research that combines the real world and computer-generated data (Ratajczak *et al.*, 2019; Akram, 2019; Webster *et al.*, 1996). Fundamentally, it is an environment where data generated by a computer is inserted into the user's view of a real-world scene (Azuma, 1997). AR allows the user to see the real world with virtual objects superimposed or combined with them. So, augmented reality complements reality rather than eliminating it entirely (Bouchlaghem *et al.*, 1996). Integrating BIM with new approaches to accessing information, such as augmented reality (AR) which would provide facility managers with an intuitive and simple approach to communicating with the information they need from the BIM models (Becerik-Gerber *et al.*, 2012). Facility Managers are often expected to connect physical objects to database information. It makes AR a good candidate to support facility managers with their routine tasks, as their live view of space could now be complemented by the database information they need, all in one graphical user interface (Baek *et al.*, 2019; Diao and Shih, 2019; Wagner and Schmalstieg, 2003). The building information model uses information in a critical, intelligent, and context-sensitive way to integrate real-world information through augmented reality (Diao and Shih, 2019; Hyoung and Dunston, 2008). Building information with a high level of details is clearly transmitted and transported to the site by augmented reality (Mirshokraeiet, 2019; Leinonen and Kahkönen, 2012). 3D reinforcement detailing reduces the time required to modify drawings. This saves time for all project participants. Accordingly, project control time and changes during construction are significantly reduced (Ratajczak *et al.*, 2019; Mirshokraeiet, 2019; Hou and Wang, 2013). Augmented reality technology allows for more advanced and concrete implementation of construction management in the early stages of a construction project. This helps identify and solve potential problems before actual building construction (Ratajczak *et al.*, 2019; Diao and Shih, 2019; Irizarry, 2013). The integrated approach to the design and construction of construction projects have been the focus of most construction researches. Sector-oriented studies can be classified into two groups – The integration of systems or toolsets that enable both design and construction processes, and design and construction processes. Augmented reality plays a key role in both groups (Alizadehsalehi *et al.*, 2020; Ratajczak *et al.*, 2019; Moum, 2010). An integrated BIM system can fully support facility-management practices for facility owners (Baek *et al.*, 2019; Diao and Shih, 2019; Milgram and Kishino, 1994). While BIM is providing static and predefined data, the augmented reality can be used for real-time visualization and monitoring of activities and tasks. Ensuring the integration of the building information model with augmented reality can provide a platform for effectively interacting and using the data contained in a building information model with the facility-management team and subcontractors (Baek *et al.*, 2019; Diao and Shih, 2019; Dunston *et al.*, 2003).

5.3.3 Mixed Reality

Mixed reality (MR) is the hybrid of real-world and virtual environments merged together for real-time interaction of physical and digital objects. Mixed reality enables designed imagery to interact with real physical existing to some extent on a digital reality platform. MR allows both virtual design and real-world content to react to each other in real-time (Vasilevski and Birt, 2020; Wu *et al.*, 2019; Cheng *et al.*, 2020).

5.3.4 Extended Reality

Extended reality (XR) is a relatively new technology among other immersive technologies, which offers improvement opportunities for activity implementation. However, there are obstacles to XR implementation (Fig. 5.2). Extended reality contains virtual and real-life environments on the same platform allowing human-machine interaction

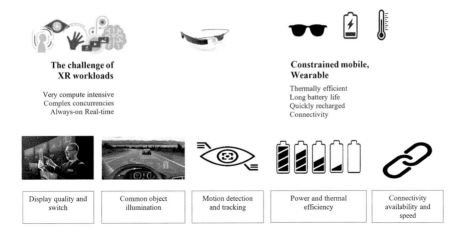

Fig. 5.2: XR technology challenges

through wearable and platform-based computer technologies. Extended reality is the unification of all AR, VR, and MR technology features on the same platform (Tromp *et al.*, 2020; Razkenari *et al.*, 2019).

In brief, augmented reality refers to creating virtual content overlaying the real-world content, which cannot allow interaction capability; virtual reality is a completely virtual environment; mixed reality is the merged version of the real-world and virtual environment in which the virtual contents can interact with the real-world environment; extended reality is a combination of three immersive technologies of AR, VR, and MR, which has a wide variety and a lot number of levels in the virtuality (Fig. 5.3).

5.4 Immersive Technology Use in the AECO/FM Industry

5.4.1 The Advantages of ImTech Use

Immersive technologies have the potential to significantly nurture the AECO/FM industry. The added value of the use of immersive technology has not yet been well understood by the industry stakeholders (Table 5.1). Moreover, immersive technology implementation is a costly method in the AECO/FM industry activities and requires time because of the learning curve. However, the advantages compared to traditional processes may change the way how the construction project processes and procedures are held. Since immersive technologies are a good method to showcase the design, the design team is the pioneer to utilize the technology most. Augmented reality may change the design of the buildings and the design approach of the design team with its intended capabilities. The

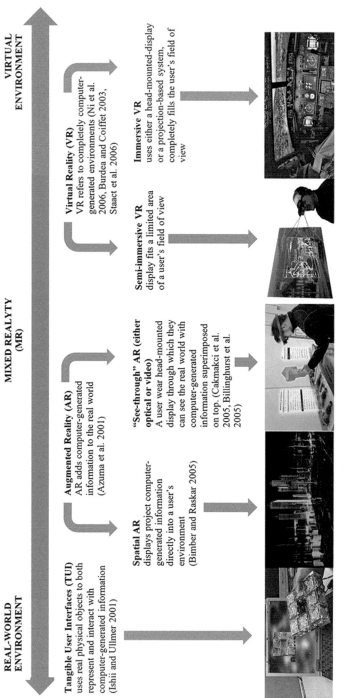

Fig. 5.3: The flow from real-world environment to virtual environment

designer may have an opportunity to sense the surrounding environment by utilizing virtual reality. The design itself is executed in the eye of the designer with the help of augmented reality glasses. Hereby, the designer experiences the design within the real environment first hand. The design team and the other stakeholders in the design phase start to cooperate and collaborate over the virtual reality representation for interchanging ideas over the design. So, the requirements of related stakeholders are discussed during the design phase without spending much time and cost for the changes in the latter phases. This results in robust conflict management by eliminating disagreements in the early phases of the project. The design failures are realized and removed during the design phase by using 3D representations, like a hologram in the person's fore-view. The design corrections in the early phase of a construction project result in efficient processes and high performance. The immersive technologies are useful in all phases, as in the design phase of a construction project. Augmented reality changes the processes and procedures by eliminating unnecessary steps to ensure lean production in the AECO/FM industry. Early interaction opportunity prevents unnecessary cost and time spent on conflicts, failures, and changes.

The contractor and construction manager's involvement in the early stages prevents conflict and loss of resources. Execution of the construction activities can be simulated for better time management. Pre-identified equipment logistics, potential conflicts, and safety violations may result in efficient processes. The construction workers can also easily have access to the blueprints with a 3D AR headset, which gives the workers the opportunity to clearly understand the design for a flawless construction job. Smart helmets allow workers to be informed of the identified potential risks and other building construction-related information. The blueprints may also be used by inspectors to compare the designed with the as-built. This is also useful in the monitoring and controlling phase for project control engineers or project managers. Bridging the gap between the digital building and the built assures correction in the end product. A building represented by augmented reality, serving of a walk-through experience, may decrease the change in a construction project, which in turn will result in lean, efficient, and cost-and-time-effective results in the AECO/FM industry. Moreover, a smart helmet can also inform the controllers about the excessive condition of pressure, blockage, potential leaking, or other potential failures that cannot be detected by a naked eye. This technology can be utilized throughout the building lifecycle by the facility managers for evaluating and feeding data visualization-related to operation and maintenance issues according to the inventory-recognition data that is integrated into the virtual model.

5.4.2 ImTech Implementation Areas and Obstacles

One of the uses of virtual reality opportunities in the AECO/FM industry is to assess risk. The code compliance can be checked easily with the effective model interaction in the reality platform. The walkthrough experience may allow code consultants or inspectors to monitor and control the emergency requirements. Furthermore, the emergency situations and evacuation process can be simulated for improving emergency plans or for training the occupants.

In general, the use of immersive technologies in the AECO/FM industry can be classified as follows:

- Design team and user interaction
- Design and construction team interaction
- Early conflict resolution
- Lean construction
- Risk assessment
- Workers' safety training
- Real-time monitoring
- Better presentation
- Marketing and sales improvement
- Design improvement
- Fast construction
- Resource management
- Efficient processes
- Construction management
- Design analysis

The obstacles of immersive-technology implementation in the AECO/FM industry can be classified as follows:
- Cost of implementation
- Learning curve
- Interoperability
- BIM integration
- Unautomated model transfer process

5.5 Immersive Technologies and BIM Integration

The integration of BIM and immersive technologies is envisaged to become an innate feature of visualization within the AECO/FM industry. However, research on BIM and immersive technologies for information sharing is rare. Many of the studies lie in the design and construction phase of a construction project. There are limitations for research in the immersive technology fields, such as social acceptance, registration problems, display devices use, data intake, interoperability, real and virtual data alignment, and capability of information-processing devices. The future of research on immersive technology seems to focus on wearable devices, construction progress monitoring, localization, speed, implementation, remote servers, and improvements in visualization. Current research in immersive technologies analyzes monitoring, inspection, education, training, and as-built data intake which has the potential for quality and efficiency improvement (Schlueter and Geyer, 2018). The integration of immersive technologies in 4D is possible for progress monitoring in order to detect schedule fallouts earlier. There are other investigations on integrating immersive technologies in 5D with visual processing and augmentation through acquired visual context (Bosch-Sijtsema *et al.*, 2019; Guo *et al.*, 2017).

5.5.1 BIM-to-VR and BIM-to-MR Implementation

VR allows all project stakeholders to virtually experience the various design and construction alternatives of a building or structure (Sidani *et al.*, 2019; Brioso *et al.*, 2019; Zhang *et al.*, 2019), thus allowing the comparison of design and construction scenarios for robust decision making of the optimal alternative in terms of design, process, material, etc. Figure 5.4

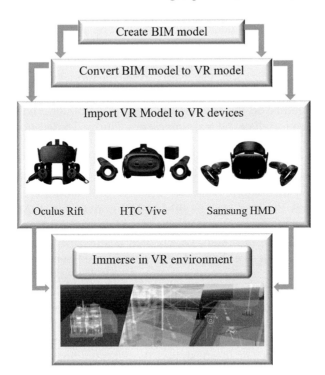

Fig. 5.4: BIM to VR integration method (adapted from Alizadehsalehi *et al.*, 2020)

presents the BIM-to-VR integration workflow, which consists of steps for creating a BIM model by BIM software, BIM model conversion to VR environment, VR model import to VR devices, and immersion in VR environment. The BIM model created by BIM software is equipped with intensive nD information, which allows the user to perceive the model from many dimensions with a parametric modeling method. A cloud-based server/database can be used to share the model online to provide real-time information to all the stakeholders. Plug in on a BIM software can enhance the conversion of the BIM model in the VR environment. The VR model in a VR environment can be accessible via VR devices, such as a head-mounted display (HMD) and a computer-aided virtual environment (CAVE).

Immersive BIM experience has advantages in terms of offering all stakeholders to examine multiple design alternatives; design decision development, evaluation, and improvement; error or clash detection, change management, quick change response, accurate information sharing, realistic design perception, improved presentation, and so on. These advantages of immersive BIM experience allow the project team to understand well the project's weaknesses and strengths, which trigger detailed improvement plans, priority list preparation, acceleration in activity realization, rapid decision-making process for more efficient and productive project results.

The real environment is allowed to be experienced by all the teams via the created MR visualization. The team can see the model and each other at the same time. Hologram technology is activated to meet these virtual and real objects together on the site. This opportunity obtains a collaborative design platform for all of the team members. Figure 5.5 demonstrates the BIM to MR conversion workflow, which consists of the steps in creating a BIM model by BIM software, BIM model conversion to MR environment, MR model upload to the cloud system, MR model import to MR devices, and immersion in MR environment. The nD BIM model creation with BIM software is the first step in BIM-to-MR conversion. BIM model is converted to the MR model via APIs, such as HoloLive, 3D Viewer, Fuzor, and HoloView in the next step, which then is stored in a cloud system. Online model storing is enhanced by a cloud-based server/database to provide real-time accessible information to all the stakeholders from anywhere with an internet connection. The MR devices' and teams' information exchange and communication capability are increased by cloud-computing. MR devices are used to offer MR experience to users via BIM model projection over a real-world environment, which serves as a projected display of the designed model, allowing immersive and interactive team collaboration.

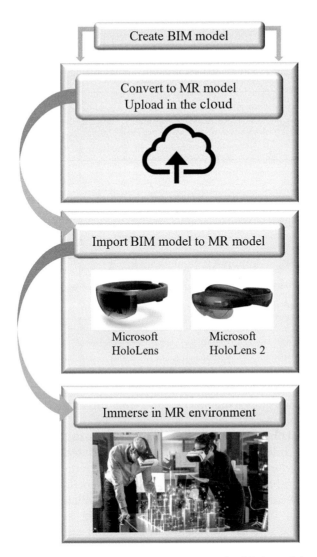

Fig. 5.5: BIM to MR integration method (adapted from Alizadehsalehi *et al.*, 2020)

5.5.2 BIM-to-XR Model Implementation

BIM-to-XR conversion is used to understand complex projects and the conversion activities are presented in Fig. 5.6 (Adhami *et al.*, 2019; Alizadehsalehi *et al.*, 2020; Waissi *et al.*, 2015). The boxes represent the key steps in the process. There are five major steps in the BIM-to-XR conversion process – the model creation, schedule creation, XR model

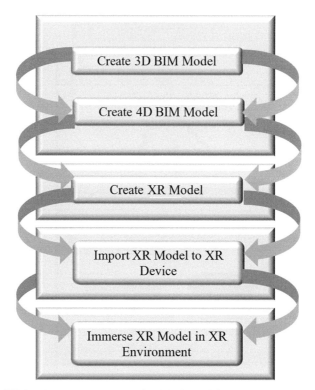

Fig. 5.6: BIM to XR integration method (adapted from Alizadehsalehi *et al.*, 2020)

creation, XR model import to XR devices, and immerse in XR environment. Design data, schedule data, cost data, and other related data about the project are gathered in the beginning. The integrated data model as a BIM model is created and the XR model is determined according to this model information with additional specifications and collaborative design-supporting team's ideas. The fully integrated and improved model alternatives are ready to be accessed and evaluated via XR devices for better project results.

5.5.3 IM-to-Game Engine Model Implementation

Nowadays, the game engine subject is a high-in-demand area of interest by various field professionals, who have less knowledge in programming but are eager to design games. Therefore, game engines serve professionals templates to easily design a game and minimize the need for programming knowledge for developing a game and exporting it to various platforms. There are many worldwide popular game engines, such as: GameMaker, JMonkey, Marmalade, the OGRE 3D, the Shiva, the Sio2, the Turbulenz, the Unity, Unreal.

Visualization is one of the vital methods to improve and expand AECO/FM activities throughout the project lifecycle. A proper visual representation of the designed or built entity and its environment serves all stakeholders for various purposes. There is an accelerated demand on interoperable platforms to ease visual representation for data exchange with different formats. There are several different data formats used in the AECO/FM industry. Game engines are more feasible and serve a wide range of fields. The import and conversion of various data formats into mesh objects by various stakeholders are faster and workable with game engines. This feature ensures a high level of collaboration and interaction between stakeholders. Visualization of a building, from initiation to demolition, is an important source of information for all AECO/FM activities utilized by decision-makers and actualizers. BIM, as an nD information platform, is widely used in the AECO/FM industry. However, the visual representation of the BIM model in a game engine platform to a virtual reality environment provides many more opportunities and advantages (Asgari and Pour, 2017). The BIM and game engine integration is utilized in as-designed and as-built BIM model view in the virtual environment; changes to design tracking, monitoring, and control (Du *et al.*, 2018b; Johansson, 2016); collaborative design with immediate inputs of stakeholders; well-informed and collaborative decision making (Du *et al.*, 2018a; Iris, 2019; Enscape, 2019); remote work, site visit, design, control; staff training for safety (Li *et al.*, 2012; Sachs *et al.*, 2013); design communication (Protchenko *et al.*, 2018; Wang *et al.*, 2014); user interaction and so on.

The implementation of the BIM model into a game engine platform needs several steps and conversions. Current BIM software packages facilitate direct import of the BIM model into the game engine. There may be data loss during this process (Protchenko *et al.*, 2018). There are several methods for the BIM model into import of game engine (Protchenko *et al.*, 2018; Johansson, 2016; Wang *et al.*, 2014; Motamedi *et al.*, 2017). Additionally, point cloud data can be imported in BIM packages while cannot be directly imported in game engines without any data loss (Stets *et al.*, 2017; Bruder *et al.*, 2014; Berge *et al.*, 2016; De Lima Hermandez *et al.*, 2019; Discher *et al.*, 2018; Thiel *et al.*, 2018). Mesh models, a derivate product of point clouds, are frequently implemented in heritage projects (Anderson *et al.*, 2010; Carrozzino and Bergamasco, 2010; Rua and Alvito, 2011; Bassier *et al.*, 2018). There are very rare implementations of mesh models in the AECO/FM industry (Kim *et al.*, 2018; Gheisari *et al.*, 2016; Wang *et al.*, 2014; Zollmann *et al.*, 2014; Kim and Kano, 2008). Apart from design and construction monitoring, progress and discrepancy checking, controlling, decision making, visualizing and other common uses of game engine integration, qualitative assessment of construction elements is realized by Kwon *et al* (2014) without any positioning feature in the architecture of the software and by Zhou *et al.* (2017) with a location-based correct placement of construction elements feature on the software's architecture.

A common format is needed for robust data conversion, share, and reuse in different referencing systems. Otherwise, data loss, inaccuracy or misinforming can occur during or after the sharing or import process (Vincke and Vergauwen, 2019). Therefore, data format conversion is needed for interoperability. There are alternative methods to transfer BIM model data to game engine platform with a data format conversion process (Fig. 5.7). Visualization of the BIM model on a game engine platform allows model access to all stakeholders, including project managers, designers, and even the client. A stand-alone model on a game engine platform eliminates the need for expensive software, need to visualize and control the design, and the project progress.

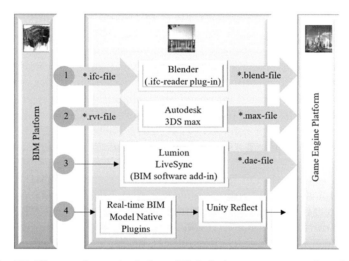

Fig. 5.7: File transfer methods from BIM platform to game engine platform (adapted from Vincke *et al.*, 2019)

5.6 Digital Twin

5.6.1 The Evolution of Digital Technology

Industrial revolutions eased and developed the production processes throughout history. Increasing product rates, project size, and project complexity led to improvements in production processes. The computational capability of the industries developed with web technologies of Web 1.0, Web 2.0, Web 3.0, and Web 4.0 respectively. The human-computer interaction enables better collaboration through dynamic knowledge generation and institutionalized knowledge that is open, transparent, fast, accurate, and reliable, will be in action for full integration among 'things' and cyber technologies (Quinn *et al*. 2020; Ozturk and Yitmen, 2019, Puyan *et al*., 2019; Alves *et al*., 2017). The new information era in the industry replaced the analog, mechanic, and electronic technology with digital technologies via improvements in information communication technologies (ICT) (Ozturk and Yitmen, 2019). With the emergence of Industry 4.0 and third generation ICT, digital technologies have become active in the value proposition of improving efficiency (Gamil *et al*., 2020; Herterich *et al*., 2016). Therefore, digital technologies may enhance risk mitigation, automated manufacturing systems, market intelligence, and service innovation via in-service information acquisition, retrieval, and store; and knowledge creation, use, share, and regeneration (Ozturk and Yitmen, 2019; Rymaszewska *et al*., 2017). Despite the slow adaptation speed of the AECO/FM industry, digitalization improved task realization, communication, process efficiency, and information utilization via interoperable and transparent information for decentralized and automated decision-making (Habibi, 2017). The new era in the AECO/FM industry results in smart buildings, smart city, business web, smart logistics, smart grid, smart mobility in production, and so on. Interoperability is the key to robust implementation of digital systems. Interacting intelligent systems improve robust knowledge utilization for better systems and processes, better and automated decision making, improved safety, improved data communication, increased production, decreased unnecessary activities, etc. (Tsilimantou *et al*., 2020; Ozturk and Yitmen, 2019). The digitalization of the AECO/FM industry will revolutionize planning, design, construction, operation, and facility management of structures.

Emergence and development of new digital technologies (Digital Twin, immersive technologies, Internet of Things, machine learning, and so on enables new service approaches in many industries (Atlam and Wills, 2020; Cheng *et al*., 2020; Carmona *et al*., 2019). Digital Twin (DT), as one of the digital technology platforms, is very promising with the virtual representation of physical assets (Atlam and Wills, 2020; Tao *et al*., 2018; Herterich *et al*., 2016). The Digital Twin idea was first coined by Grieves in 2003 as a part of a Product Lifecycle Management (PLM) concept. This approach was a 'conceptual ideal for PLM'. Then came the idea called 'Information Mirroring Model' by Grieves in 2006. The final evolution of the name of the concept was stated as 'Digital Twin' in 2010 (Grieves, 2014). The Digital Twin concept turned from being a PLM tool to a digital platform (Tao *et al*., 2018). This evolutionary path resulted in two main characteristics of the concept – the full product lifecycle integration and up-to-date dynamic data generation via cognitive technologies. Digital Twin equipped with sensors, gauges, measuring machines, lasers, vision systems, and white light scanning can sense the real-life experience information of the physical asset, so that the Digital Twin can accurately predict the possible failures, feed information back to the system, and react according to the stimulant information (Tao *et al.*, 2018; Herterich *et al.*, 2016). Digital Twin provides the opportunity to acquire information about the physical asset throughout its lifecycle and develop all project processes continuously (Atlam and Wills, 2020; Tao *et al.*, 2018).

The developments in technology facilitate high amount and quality of information integrated into the virtual and physical products. The feedback and feed-forward loops of knowledge from physical products to virtual ones or vice versa cannot be implemented smoothly yet. The simultaneous interaction between physical and virtual products should be fully achieved for fulfilling the promise of the Digital Twin concept.

5.6.2 Benefits of Digital Twin Adoption

The benefits of Digital Twin adoption in the AECO/FM industry are reduced construction and operating costs, increased productivity, increased collaboration, improved safety, optimized performance, and sustainability. Virtual construction scenarios familiarize the construction team with tasks, resulting in a decrease in change and rework, which in turn reduce cost and time overruns. Artificial intelligence, machine learning, and data-driven decision making enhance prediction in construction activities while expediting the unexpected costs throughout the project lifecycle. The up-to-date design, material specification, schedule, and inventories information of the building can be stored, updated, accessed, analyzed, and used throughout the lifecycle. Real-time site tracking information share can be available with the site team for alerts and notifications in case of emergency situations, like monitoring and analysis of facility performance over real-time operational data. Operation and maintenance optimization can be enabled via facility monitoring and analysis.

The physical entity and the Digital Twin are connected via data platforms and aggregators. The interaction of these two counterparts enhances data integration for many afore-mentioned reasons. The early development of Digital Twin results in effective outcomes and can be developed at any stage of the building lifecycle. The maturity of a Digital Twin increases with the data integrated into it. Six levels are identified for Digital Twin maturity levels. Any Digital Twin can adopt any maturity level free from the maturity level sequence. The maturity levels are namely reality capture, 3D model, static data connection to the model, real-time data integration, data integration and interaction, and autonomous operations and maintenance. Respectively, the levels consist of as-built virtual data capture via drones, drawings, point clouds, or photogrammetry. The object-based design optimization and coordination are realized over the 3D model. The static data connection to the 3D model is enhanced with 4D/5D data integration for design management by documents, asset management systems, and drawings. Operational efficiency is increased by real-time data integration via IoT sensors. Bidirectional data integration and interaction can be possible with remote and immersive operations to control the physical world. Self-governance of the building with autonomous operations and maintenance can be possible with total oversight and transparency. The data will be utilized by data consumers and stakeholders with data-access devices, tools, and APIs while inbreeding the data to the ecosystem through a data platform.

The fourth maturity level of the Digital Twin is related to immersive technologies. The Digital Twin and physical counterpart integration and interaction can be realized with information feedback and feed-forward (Fig. 5.8). The level of maturity requires sensor and immersive data. The Digital Twin or its physical counterpart can feed information in two-ways – interaction and communication type. The immersive technology integration allows human-to-machine and machine-to-human interaction over the full integration of both digital and physical assets. A Digital Twin of a physical object can be generated via immersive technologies to display as hologram via use of immersive technology devices, thereby the offsite interaction with the physical entity can be possible with human-to-computer interaction or vice versa.

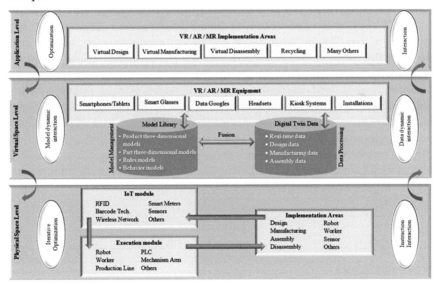

Fig. 5.8: The application framework of VR, AR, and MR in Digital Twin

5.7 Conclusion

BIM and ImTech integration (AR, VR, MR, XR) have proven its potential in revolutionizing design, construction, operation, and facility-management activities in the industry. Effective integration of these technologies can enhance information generation, store, share, use, and management. Research in ImTech and BIM integration increasing, however, needs more attention. This chapter presents methods, process steps, software, and devices for ImTech and BIM integration. AECO/FM industry can benefit from utilizing ImTech and BIM integration in improving time, cost, quality, and safety performance. This integration may be effective throughout all phases, ranging from initiation to facility management and all activities in design to maintenance. In this chapter, technological, content, and user challenges, and solutions for the ImTech and BIM integration are explained, based on existing literature with potential applications in design management, safety management, progress monitoring, conflict avoidance, decision-making, and other subjects. Content generation, transfer, and delivering of the content are the fundamental challenges in integration. The ImTech and BIM technology and its related hardware and software systems are constantly updated. The users may face difficulties to access these updates in the power of purchase. Besides the adaptation process to learn and accept the newest systems is another challenge for users. The agility of the AECO/FM industry to overcome these challenges will determine the future of the subject in the construction field. However, it is inevitable that ImTech and BIM integration will progressively change the AECO/FM industry. Moreover, artificial intelligence, machine learning, the Internet of Things, and other cognitive technology integration will increase the effect of technology in AECO/FM industry. The technology improvements in bandwidth, speed, 5G wireless connection, and data transfer speeds will increase online collaboration and efficient work environment within automated and improved processes. Future works will focus on improving the technologies for integration and integration process for ease, speed, and efficiency in BIM to ImTech and in ImTech to BIM conversions. Further research is needed in integrating other cognitive technologies for more mature BIM models. Investigation on all dimensions and all types, sizes, and phases of projects for integration is needed in the near future.

References

Abbas, A., Choi, M., Seo, J., Cha, S.H. and Li, H. (2019). Effectiveness of immersive virtual reality-based communication for construction projects. *KSCE J. Civ. Eng.*, 23: 4972-4983. https://doi.org/10.1007/s12205-019-0898-0

Akinci, B., Boukamp, F., Gordon, C., Huber, D., Lyons, C. and Park, K. (2006). A formalism for utilization of sensor systems and integrated project models for active construction quality control. *Autom. Constr.*, 15(2): 124-138. https://doi.org/10.1016/j.autcon.2005.01.008

Akram, R., Thaheem, M.J., Nasir, A.R., Ali, T.H. and Khan, S. (2019). Exploring the role of building information modeling in construction safety through science mapping. *Safety Science*, 120: 456-470. https://doi.org/10.1016/j.ssci.2019.07.036

Al-Adhami, M., Wu, S. and Ma, L. (2019). Extended Reality Approach for Construction Quality Control. *CIB World Building Congress 2019*, Hong Kong SAR, China 1721 June 2019.

Alam, M.F., Katsikas, S. and Beltramello, O. (2017). Hadjiefthymiades, S. Augmented and virtual reality-based monitoring and safety system: A prototype IoT platform. *J. Netw. Comput. Appl.*, 89: 109-119. https://doi.org/10.1016/j.jnca.2017.03.022

Alizadehsalehi, S., Hadavi, A. and Huang, J.C. (2019a). Virtual reality for design and construction education environment. *AEI 2019*, American Society of Civil Engineers, Reston, VA, (2019a): 193-203. doi: 10.1061/9780784482261.023

Alizadehsalehi, S., Hadavi, A. and Huang, J.C. (2019b). BIM/MR-lean construction project delivery management system (2019b). IEEE Technology & Engineering Management Conference (TEMSCON) *IEEE*: 1-6. doi: 10.1109/TEMSCON.2019.8813574

Alizadehsalehi, S., Hadavi, A. and Huang, J.C. (2020). From BIM to extended reality in AEC industry. *Automation in Construction*, 116: 103254. DOI: 10.1016/j.autcon.2020.103254

Alves, M., Carreira, P. and Costa, A.A. (2017) BIMSL: A generic approach to the integration of building information models with real-time sensor data. *Automation in Construction*, 304-314. DOI:10.1016/j.autcon.2017.09.005

Ammari, K.E. and Hammad, A. (2014). Collaborative BIM-based markerless mixed reality framework for facilities maintenance. *In*: Proceedings of the 2014 International Conference on Computing in Civil and Building Engineering, Orlando, FL, USA. 23-25 June 2014; pp. 657-664. https://doi.org/10.1061/9780784413616.082

Anderson, E.F., McLoughlin, L., Liarokapis, F., Peters, C., Petridis, P. and de Freitas, S. (2010). Developing serious games for cultural heritage: A state-of-the-art review. *Virtual Reality*, 14(4): 255-275. https://doi.org/10.1007/s10055-010-0177-3

Anil, E.B., Tang, P., Akinci, B. and Huber, D. (2013). Deviation analysis method for the assessment of the quality of the as-is Building Information Models generated from point cloud data, *Autom. Constr.*, 35: 507-516. https://doi.org/10.1016/j.autcon.2013.06.003

Asgari, Z. and Pour, F. (2017). Advanced virtual reality applications and intelligent agents for construction process optimisation and defect prevention. *Procedia Engineering*, 196: 1130-1137. https://doi.org/10.1016/j.proeng.2017.08.070

Atlam, H.F. and Wills, G.B. (2020). IoT security, privacy, safety and ethics. *In*: Farsi, M., Daneshkhah, A., Hosseinian-Far, A., Jahankhani, H. (Eds.). Digital Twin Technologies and Smart Cities. Internet of Things (Technology, Communications and Computing), Springer, Cham. https://doi.org/10.1007/978-3-030-18732-3_8

Azhar, S. (2017). Role of visualization technologies in safety planning and management at construction jobsites. *Procedia Engineering* 171: 215-226 DOI: 10.1016/j.proeng.2017.01.329

Azuma, R.T. (1997). A survey of augmented reality. *Teleoperators and Virtual Environments, Presence*, 6(4): 355-385.

Baek, F., Ha, I. and Kim H., (2019). Augmented reality system for facility management using image-based indoor localization. *Automation in Construction*, 99: 18-26. https://doi.org/10.1016/j.autcon.2018.11.034

Bae, H., Golparvar-Fard, M. and White, J. (2013). High-precision vision-based mobile augmented reality system for context-aware architectural, engineering, construction and facility management (AEC/FM) applications. *Visualization in Engineering*, 1(1): 3. https://doi.org/10.1186/2213-7459-1-3

Bassier, M., Vincke, S., Hernandez, R.D.L. and Vergauwen, M. (2018). An overview of innovative heritage deliverables based on remote sensing techniques. *Remote Sensing*, 10(10): 1607. DOI: 10.3390/rs10101607

Becerik-Gerber, B., Farrokh, J., Li, N. and Calis, G. (2012). Application areas and data requirements for BIM-enabled facilities management. *Journal of Construction Engineering and Management*, 138: 431-442. doi: 10.1061/(ASCE)CO.1943-7862.0000433

Behzadan, A.H. and Kamat, V.R. (2009). Interactive augmented reality visualization for improved damage prevention and maintenance of underground infrastructure. Proceedings of the 2009 Construction Research Congress, Seattle, WA.

Berge, L.P., Aouf, N., Duval, T. and Coppin, G. (2016). Generation and VR visualization of 3D point clouds for drone target validation assisted by an operator. *In*: Proceedings CEEC 2016 : 8th Computer Science and Electronic Engineering Conference, IEEE, pp. 66-70.

Billinghurst, M. and Dünser, A. (2012). Augmented reality in the classroom. *Computer*, 45(7): 56-63.

Bosch-Sijtsema, P.M., Gluch, P. and Sezer, A.A. (2019). Professional development of the BIM actor role. *Automation in Construction*, (97): 44-51. doi: 10.1016/j.autcon.2018.10.024

Bosché, F., Ahmed, M., Turkan, Y., Haas, C.T. and Haas, R. (2015). The value of integrating Scan-to-BIM and Scan-vs-BIM techniques for construction monitoring using laser scanning and BIM: The case of cylindrical MEP components. *Autom. Constr.*, 49: Part B 201-213. https://doi.org/10.1016/j.autcon.2014.05.014.

Botella, C., Breton-López, J., Quero, S., Baños, R.M., Garcia-Palacios, A., Zaragoza, I. and Alcaniz, M. (2011). Treating cockroach phobia using a serious game on a mobile phone and augmented reality exposure: A single case study. *Computers in Human Behavior*, 27(1): 217-227.

Bouchlaghem, N.M. and Liyanage, I.G. (1996). Virtual reality applications in the UK's construction industry. CIB W78 Conference, Construction on the Information Highway. University of Ljubljana, Slovenia., pp. 89-94.

Brioso, X., Calderón, C., Aguilar, R. and Pando, M.A. (2019). Preliminary methodology for the integration of lean construction BIM and virtual reality in the planning phase of structural intervention in heritage structures. pp. 484-492. *In:* Structural Analysis of Historical Constructions, Springer.

Brown, R. and Barros, A. (2013). Towards a service framework for remote sales support via augmented reality. Web Information Systems Engineering – WISE 2011 and 2012 Workshops. Springer, Berlin Heidelberg, pp. 335-347.

Bruder, G., Steinicke, F. and Andreas, N. (2014). Poster: Immersive point cloud virtual environments. *In*: IEEE Symposium on 3D User Interfaces, pp. 161-162.

Burke, J.W., McNeill, M., Charles, D., Morrow, P.J., Crosbie, J. and McDonough, S. (2010). Augmented reality games for upper-limb stroke rehabilitation. 2010 Second International Conference on Games and Virtual Worlds for Serious Applications (VS-GAMES), IEEE, Braga.

Cabero-Almenara, J. and Roig-Vila, R. (2019). The motivation of technological scenarios in augmented reality (AR): Results of different experiments. *Appl. Sci.*, 9: 2907.

Carmona, A.M., Chaparro, A.I., Velasquez, R., Botero-Valencia, J., Castano-Londono, L., Marquez-Viloria, D. and Mesa, A.M. (2019). Instrumentation and data collection methodology to enhance productivity in construction sites using embedded systems and IoT technologies. *In*: Mutis, I. and Hartmann, T. (Eds.). Advances in Informatics and Computing in Civil and Construction Engineering. Springer, Cham. https://doi.org/10.1007/978-3-030-00220-6_76

Chalhoub, J., Alsafouri, S. and Ayer, S.K. (2018). Leveraging site survey points for mixed reality BIM visualization. Construction Research Congress 2018: 326-335. DOI: 10.1061/9780784481264.032

Chalhoub, J. and Ayer, S.K. (2018). Using mixed reality for electrical construction design Communication. *Automation in Construction*, (86): 1-10. doi: 10.1016/j.autcon.2017.10.028

Chen, H.T., Wu, S.W. and Hsieh, S.H. (2013). Visualization of CCTV coverage in public building space using BIM technology. *Visualization in Engineering*, 1(1): 1-17.

Cheng, J.C.P., Chen, K. and Chen, W. (2020). State-of-the-art review on mixed reality applications in the AECO industry. *Journal of Construction Engineering and Management*, 146(2). https://doi.org/10.1061/(ASCE)CO.1943-7862.0001749

Cheng, J.C.P., Chen, W., Chen, K. and Wang, Q. (2020). Data-driven predictive maintenance planning framework for MEP components based on BIM and IoT using machine learning algorithms. *Automation in Construction*, 112: 03087. https://doi.org/10.1016/j.autcon.2020.103087

Chu, M., Matthews, J. and Love, P.E. (2018). Integrating mobile building information modelling and augmented reality systems: An experimental study. *Automation in Construction*, 85: 305-316. DOI: 10.1016/j.autcon.2017.10.032

Carrozzino, M. and Bergamasco, M. (2010). Beyond virtual museums: Experiencing immersive virtual reality in real museums. *Journal of Cultural Heritage*, 11(4): 452-458. https://doi.org/10.1016/j.culher.2010.04.001

Dainty, A., Moore, D. and Murray, M. (2006). *Communication in Construction: Theory and Practice*. Taylor & Francis: New York, NY, USA, pp. 19-52. ISBN 9780203358641

Davies, R. and Harty, C. (2013). Implementing 'Site BIM': A case study of ICT innovation on a large hospital project. *Autom. Constr.*, 30: 15-24. https://doi.org/10.1016/j.autcon.2012.11.024

De Lima Hernandez, R., Vincke, S., Bassier, M., Mattheuwsen, L., Derdaele, J. and Vergauwen, M. (2019). Puzzling engine: A digital platform to aid the reassembling of fractured fragments. To be published in: International Archives of the Photogrammetry, Remote Sensing and Spatial Information Sciences Photogrammetry, CIPA 2019, paper submitted for publication (June 2019).

Diao, P.H. and Shih, N.J. (2019). BIM-based AR maintenance system (BARMS) as an intelligent instruction platform for complex plumbing facilities. *Appl. Sci.*, 9: 1592.

Discher, S., Masopust, L., Schulz, S., Richter, R. and Dollner, J. (2018). A point-based and image-based multi-pass rendering technique for visualizing massive 3D point clouds in VR environments. *In*: 26th International Conference on Computer Graphics, Visualization and Computer Vision.

Du, J., Zou, Z., Shi, Y. and Zhao, D. (2017). Simultaneous data exchange between BIM and VR for collaborative decision making. *Computing in Civil Engineering 2017*, 1-8. doi: 10.1061/9780784480830.001

Du, J., Shi, Y., Zou, Z. and Zhao, D. (2018a). CoVR: Cloud-based multiuser virtual reality headset system for project communication of remote users. *Journal of Construction Engineering and Management*, 144(2): 1-19. https://doi.org/10.1061/(ASCE)CO.1943-7862.0001426

Du, J., Zou, Z., Shi, Y. and Zhao, D. (2018b). Zero latency: Real-time synchronization of BIM data in virtual reality for collaborative decision-making. *Automation in Construction*, (85): 51-64. doi: 10.1016/j.autcon.2017.10.009

Dunston, P.S., Wang, X., Billinghurst, M. and Hampson, B. (2003). *Mixed Reality Benefits for Design Perception*, 1-16.

Dunston, P.S. and Wang, X. (2005). Mixed reality-based visualization interfaces for architecture, engineering and construction industry, ASCE. *J. Constr. Eng. Manag.*, 131(12): 1301-1309. http://dx.doi.org/10.1061/(asce)0733-9346, 131:12(1301)

Enscape (2019). Enscape. https//enscape3d.com.

Fiorentino, M., Uva, A.E., Gattullo, M., Debernardis, S. and Monno, G. (2014). Augmented reality on large screen for interactive maintenance instructions. *Comput. Ind.*, 65: 270-278. https://doi.org/10.1016/j.compind.2013.11.004

Froehlich, M.A. and Azhar, S. (2016). Investigating virtual reality headset applications in construction. Proceedings of the 52nd ASC International Conference, Provo, UT, http://ascpro0.ascweb.org/archives/cd/2016/paper/CPRT195002016.pdf

Gamil, Y., Abdullah, M.A., Rahman, I.A. and Asad, M.M. (2020). Internet of things in construction industry revolution 4.0: Recent trends and challenges in the Malaysian context. *Journal of Engineering, Design and Technology*. Ahead of print. https://doi.org/10.1108/JEDT-06-2019-0164

Getuli, V., Capone, P., Bruttini, A. and Isaac, S. (2020). BIM-based immersive virtual reality for construction workspace planning: A safety-oriented approach. *Automation in Construction*, 114: 103160. https://doi.org/10.1016/j.autcon.2020.103160

Gheisari, M., Sabzevar, M.F., Chen, P. and Irizzary, J. (2016). Integrating BIM and panorama to create a semi-augmented-reality experience of a construction site. *International Journal of Construction Education and Research*, 12(4): 303-316. https://doi.org/10.1080/15578771.2016.1240117

Goldman, Sachs (2016). The real deal with virtual and augmented reality. Accessed July 10, 2020. https://www.goldmansachs.com/insights /pages/virtual-and-augmented-reality.html

Golparvar-Fard, M., Pen-a-Mora, F. and Savarese, S. (2011). Integrated sequential as-built and as-planned representation with D4AR tools in support of decision-making tasks in the AEC/FM industry. *J. Constr. Eng. Manag.*, 137: 1099-1116. https://doi.org/10.1061/(ASCE)CO.1943-7862.0000371

Grieves, M. (2014). Digital Twin: Manufacturing Excellence through Virtual Factory Replication, white paper, 1: 1-7.

Guo, H., Yu, Y. and Skitmore, M. (2017). Visualization technology-based construction safety management: A review. *Autom. Constr.*, 73: 135-144. https://doi.org/10.1016/j.autcon.2016.10.004

Habibi, S. (2017). Micro-climatization and real-time digitalization effects on energy efficiency based on user behavior. *Building and Environment*, 114: 410-428. DOI: 10.1016/j.buildenv.2016.12.039

Hagedorn, B. and Döllner, J. (2007). High-level web service for 3D building information visualization and analysis. Proceedings of the 15th Annual ACM International Symposium on Advances in Geographic Information Systems, ACM, New York, NY.

Haggard, K.E. (2017). Case Study on Virtual Reality in Construction. https://digitalcommons.calpoly.edu/cmsp/54

Han, S., Achar, M., Lee, S. and Peña-Mora, F. (2013). Empirical assessment of a RGB-D sensor on motion capture and action recognition for construction worker monitoring. *Visualization in Engineering*, 1(1): 1-13.

Herterich, M.M., Eck, A. and Uebernickel, F. (2016). Exploring how digitized products enable industrial service innovation. 24th European Conference on Information Systems, 1-17.

Hou, L. and Wang, X. (2013). A study on the benefits of augmented reality in retaining working memory in assembly tasks: A focus on differences in gender. *Automation in Construction*, 32: 38-45.

Hou, L., Wang, X., Bernold, L. and Love, P.E. (2013). Using animated augmented reality to cognitively guide assembly. *Journal of Computing in Civil Engineering*, 27(5): 439-451.

Hou, L., Wang, X. and Truijens, M. (2015). Using augmented reality to facilitate piping assembly: An experiment-based evaluation. *ASCE J. Comput. Civ. Eng.*, 29(1). http:// dx.doi.org/10.1061/(ASCE)CP.1943-5487.0000344, 05014007

Hsiao, K.F., Chen, N.S. and Huang, S.Y. (2012). Learning while exercising for science education in augmented reality among adolescents. *Interactive Learning Environments*, 20(4): 331-349.

Hu, W. and Guo, S. (2011). Visualization and collaboration of on-site environments based on building information model for construction project class. *Information and Management Engineering*, 235: 288-295.

Huynh, D.N.T., Raveendran, K., Xu, Y., Spreen, K. and MacIntyre, B. (2009). Art of defense: A collaborative handheld augmented reality board game. Proceedings of the 2009 ACM SIGGRAPH Symposium on Video Games, ACM, New York, NY.

Hyoung, D. and Dunston, P.S. (2008). Identification of application areas for Augmented Reality in industrial construction based on technology suitability. *Automation in Construction*, 17(7): 882-894 DOI: 10.1016/j.autcon.2008.02.012

Iris (2019). IrisVR. https//Irisvr.com/prospect.

Irizarry, J., Gheisari, M., Williams, G. and Walker, B.N. (2013). Infospot: A mobile augmented reality method for accessing building information through a situation awareness approach. *Autom. Constr.*, 33: 11-23. http://dx.doi.org/10.1016/j.autcon. 2012.09.002

Jaselskis, E.J., Anderson, M.R., Jahren, C.T., Rodriguez, Y. and Njos, S. (1995). Radio-frequency identification applications in construction industry. *J. Constr. Eng. Manag.*, 121: 189-196. https://doi.org/10.1061/(ASCE)0733-9364

Jin, Z., Gambatese, J., Liu, D. and Dharmapalan, V. (2019). Using 4D BIM to assess construction risks during the design phase. *Engineering, Construction and Architectural Management*, 26(11): 2637-2654. https://doi.org/10.1108/ECAM-09-2018-0379

Johansson, M. (2016). *From BIM to VR – The Design and Development of BIMXplorer*. Gothenburg: Chalmers University of Technology.

Johansson, M., Roupé, M. and Viklund Tallgren, M. (2014). From BIM to VR-integrating immersive visualizations in the current design process. *In*: Fusion-Proceedings of the 32nd eCAADe Conference, Volume 2 (eCAADe 2014), pp. 261-269.

Kaiser, R. and Scatsky, D. (2017). For more companies, new ways of seeing. Accessed July 10, 2020. https://www2.deloitte.com/insights/us/en /focus/signals-for-strategists/augmented-and-virtual-reality-enterprise-applications.html

Kamarainen, A.M., Metcalf, S., Grotzer, T., Browne, A., Mazzuca, D., Tutwiler, M.S. and Dede, C. (2013). EcoMOBILE: Integrating augmented reality and probeware with environmental education field trips. *Computers & Education*, 68: 545-556.

Kim, J. and Irizarry, J. (2020). Evaluating the use of augmented reality technology to improve construction management student's spatial skills. *International Journal of Construction Education and Research*, 1-18. https://doi.org/10.1080/15578771.2020.1717680

Kim, H. and Kano, N. (2008). Comparison of construction photograph and VR image in construction progress. *Automation in Construction*, 17: 137-143. DOI:10.22260/ISARC2005/0027

Kim, C. and Kim, C. (2012). An integrated system for automated construction progress visualization using IFC-Based BIM. *ISARC Proceedings*, 11(2): 77.

Kim, H.S., Kim, S.K., Borrmann, A. and Kang, L.S. (2018). Improvement of realism of 4D objects using augmented reality objects and actual images of a construction site. *KSCE Journal of Civil Engineering*, 22(8): 2735-2746. https://doi.org/10.1007/s12205-017-0734-3

Kimoto, K., Endo, K., Iwashita, S. and Fujiwara, M. (2005). The application of PDA as mobile computing system on construction management. *Autom. Constr.*, 14(4): 500-511. https://doi.org/10.1016/j.autcon.2004.09.003

Kivrak, S. and Arslan, G. (2019). Using augmented reality to facilitate construction site activities. *In*: Mutis, I., Hartmann, T. (Eds.). Advances in Informatics and Computing in Civil and Construction Engineering. Springer, Cham. https://doi.org/10.1007/978-3-030-00220-6_26

Klatzky, R.L., Wu, B., Shelton, D. and Stetten, G. (2008). Effectiveness of augmented-reality visualization versus cognitive mediation for learning actions in near space. *ACM Transactions on Applied Perception (TAP)*, 5(1): 1.

Klempous, R., Kluwak, K., Idzikowski, R., Nowobilski, T. and Zamojski, T. (2017). Possibility analysis of danger factors visualization in the construction environment based on virtual reality model. Cognitive Infocommunications (CogInfoCom), 2017 8th IEEE International Conference on IEEE, 2017, pp. 000363-000368. doi: 10. 1109/CogInfoCom.2017.8268271

Koch, C., Neges, M., König, M. and Abramovici, M. (2014). Natural markers for augmented reality-based indoor navigation and facility maintenance. *Autom. Constr.*, 48: 18-30. https://doi.org/10.1016/j.autcon.2014.08.009

Konig, M., Koch, C., Habenicht, I. and Spieckermann, S. (2012). Intelligent BIM-based construction scheduling using discrete event simulation. Simulation Conference (WSC), Proceedings of the 2012 Winter, Berlin.

Kopsida, M. and Brilakis, I. (2020). Real-time volume-to-plane comparison for mixed reality–based progress monitoring. *Journal of Computing in Civil Engineering*, 34(4). https://doi.org/10.1061/(ASCE)CP.1943-5487.0000896

Kuliga, S.F., Thrash, T., Dalton, R.C. and Hoelscher, C. (2015). Virtual reality as an empirical research tool-exploring user experience in a real building and a corresponding virtual model. *Computers, Environment and Urban Systems*, (54): 363-375. doi: 10.1016/j.compenvurbsys.2015.09.006

Kwon, O.S., Park, C.S. and Lim, C.R. (2014). A defect management system for reinforced concrete work utilizing BIM, image-matching and augmented reality. *Autom. Constr.*, 46: 74-81. https://doi.org/10.1016/j.autcon.2014.05.005

Lamounier, E., Bucioli, A., Cardoso, A., Andrade, A. and Soares, A. (2010). On the use of augmented reality techniques in learning and interpretation of cardiologic data. *Engineering in Medicine and Biology Society (EMBC)*, 2010 Annual International Conference of the IEEE, IEEE, Buenos Aires.

Lee, S. and Akin, Ö. (2011). Augmented reality-based computational fieldwork support for equipment operations and maintenance. *Autom. Constr.*, 20: 338-352. https://doi.org/10.1016/j.autcon.2010.11.004

Leinonen, J. and Kahkönen, K. (2012). New construction management practice based on the virtual reality technology. *Construction Congress VI* (s. 1-9). Orlando, Florida, United States.

Li, H., Chan, G. and Skitmore, M. (2012). Visualizing safety assessment by integrating the use of game technology. *Automation in Construction*, 22: 498-505.

Li, X., Yi, W., Chi, H.L., Wang, X. and Chan, A.P. (2018). A critical review of virtual and augmented reality (VR/AR) applications in construction safety. *Automation in Construction*, (86): 150-162. doi: 10.1016/j.autcon.2017.11.003. https://doi.org/10.1016/j.autcon.2011.11.009

Liu, R. and Issa, R. (2012). 3D visualization of sub-surface pipelines in connection with the building utilities: Integrating GIS and BIM for facility management. *Computing in Civil Engineering* (2012), ASCE, Gainesville, FL.

Liu, F. and Seipel, S. (2018). Precision study on augmented reality-based visual guidance for facility management tasks. *Autom. Constr.*, 90: 79-90. https://doi.org/10.1016/j.autcon.2018.02.020

Love, P.E.D., Edwards, D.J., Han, S. and Goh, Y.M. (2011). Design error reduction: Toward the effective utilization of building information modeling. *Res. Eng. Des.*, 22(3): 173-187. http://dx.doi.org/10.1007/s00163-011-0105-x

Ma, Z., Cai, S., Mao, N., Yang, Q., Feng, J. and Wang, P. (2018). Construction quality management based on a collaborative system using BIM and indoor positioning. *Autom. Constr.*, 92: 35-45. https://doi.org/10.1016/j.autcon.2018.03.027

Manyika, J., Ramaswamy, S., Khanna, S., Sarrazin, H., Pinkus, G., Sethupathy, G. and Yaffe, A. (2015). *Digital America: A Tale of Haves and Have-mores*. San Francisco: McKinsey Global Institute.

Matsutomo, S., Miyauchi, T., Noguchi, S. and Yamashita, H. (2012). Real-time visualization system of magnetic field utilizing augmented reality technology for education. *IEEE Transactions on Magnetics*, 48(2): 531-534.

Matthews, J., Love, P.E.D., Heinemann, S., Rumsey, C., Chandler, R. and Olatunji, O. (2015). Real-time progress monitoring: Process re-engineering with cloud-based BIM in construction. *Autom. Constr.*, 58: 38-47. http://dx.doi.org/10.1016/j.autcon.2015.07.004.

McKinsey Global Institute (2016). *Digital Europe: Pushing the Frontier, Capturing the Benefits*. McKinsey Company: New York, NY, USA, pp. 7-22.

Memarzadeh, M. and Golparvar-Fard, M. (2012). Monitoring and visualization of building construction embodied carbon footprint using DnAR-N-dimensional augmented reality models. Proceedings of the Construction Research Congress, West Lafayette, Indiana, pp. 1330-1339.

Merivirta, M.L., Mäkelä, T., Kiviniemi, M., Kähkönen, K., Sulankivi, K. and Koppinen, T. (2011). Exploitation of BIM based information displays for construction site safety communication. CIB W099 Conference, Washington, DC.

Meža, S., Turk, Z. and Dolenc, M. (2014). Component based engineering of a mobile BIM-based augmented reality system. *Autom. Constr.*, 42: 1-12. http://dx.doi.org/10. 1016/j.autcon.2014.02.011

Mhalas, A., Kassem, M., Crosbie, T. and Dawood, N. (2013). A visual energy performance assessment and decision support tool for dwellings. *Visualization in Engineering*, 1(1): 1-13.

Milgram, P. and Kishino, F. (1994). A taxonomy of mixed reality visual displays. *IEICE Transactions on Information Systems*, 12(12): 1321-1329.

Mirshokraei, M., Gaetami, C.I.D. and Migliaccio, F. (2019). A web-based BIM–AR quality management system for structural elements. *Appl. Sci.*, 9: 3984. doi:10.3390/app9193984

Mo, Y., Zhao, D., Du, J., Liu, W. and Dhara, A. (2018). Data-driven approach to scenario de- termination for VR-based construction safety training. *Construction Research Congress*, 2018: 116-125. doi: 10.1061/9780784481288.012

Moon, S., Becerik-Gerber, B. and Soibelman, L. (2019). Virtual learning for workers in robot deployed construction sites. *In*: Mutis, I., Hartmann, T. (Eds.). Advances in Informatics and Computing in Civil and Construction Engineering, Springer, Cham. https://doi.org/10.1007/978-3-030-00220-6_107

Moon, S. and Yang, B. (2010). Effective monitoring of the concrete pouring operation in an RFID-based environment. *J. Comput. Civ. Eng.*, 24(1): 108-116. https://doi.org/10.1061/(ASCE)CP.1943-5487.0000004

Motamedi, A., Wang, Z., Yabuki, N., Fukuda, T. and Michikawa, T. (2017). Signage visibility analysis and optimization system using BIM-enabled virtual reality (VR) environments. *Advanced Engineering Informatics*, 32: 248-262. https://doi.org/10.1016/j.aei.2017.03.005

Moum, A. (2010). Design team stories: Exploring interdisciplinary use of 3D object models in practice. *Automation in Construction*, 19(5): 554-596. doi: 10.1016/j.autcon.2009.11.007

Muhammad, A.A., Yitmen, I., Alizadehsalehi, S. and Celik, T. (2019). Adoption of virtual reality (VR) for site layout optimization of construction projects. *Teknik Dergi.*, 31(2). 10.18400/tekderg.423448

Neges, M., Koch, C., König, M. and Abramovici, M. (2017). Combining visual natural markers and IMU for improved AR based indoor navigation. *Adv. Eng. Inform.*, 31: 18-31. https://doi.org/10.1016/j.aei.2015.10.005

Niu, S., Pan, W. and Zhao, Y. (2016). A virtual reality integrated design approach to improving occupancy information integrity for closing the building energy performance gap. *Sustainable Cities Society*, (27): 275-286. doi: 10.1016/j.scs.2016.03.010

Nnaji, C. and Karakhan, A.A. (2020). Technologies for safety and health management in construction: Current use, implementation benefits and limitations, and adoption barriers. *Journal of Building Engineering*, 29: 101212. https://doi.org/10.1016/j.jobe.2020.101212

Olorunfemi, A., Dai, F., Tang, L. and Yoon, Y. (2018). Three-dimensional visual and collaborative environment for jobsite risk communication. *Construction Research Congress*, 2018: 345-355. doi: 10.1061/9780784481288.034

Ong, S.K. and Zhu, J. (2013). A novel maintenance system for equipment serviceability improvement. *CIRP Ann.*, 62: 39-42. https://doi.org/10.1016/j.cirp.2013.03.091

Ozturk, G.B. (2020). Interoperability in building information modeling for AECO/FM industry. *Automation in Construction*, 113: 103122. https://doi.org/10.1016/j.autcon.2020.103122

Ozturk, G.B. and Yitmen, I. (2019). Conceptual model of building information modeling usage for knowledge management in construction projects. IOP Conference Series: *Materials Science and Engineering*, 471: 022043. doi:10.1088/1757-899X/471/2/022043

Paes, D., Arantes, E. and Irizarry, J. (2017). Immersive environment for improving the understanding of architectural 3D models: Comparing user spatial perception between immersive and traditional virtual reality systems. *Automation in Construction*, (84): 292-303. doi:10.1016/j.autcon.2017.09.016

Panneta, K. (2018). *Gartner Top 10 Strategic Technology Trends for 2019*. Stamford, CT: Gartner.

Park, C.S., Lee, D.Y., Kwon, O.S. and Wang, X. (2013). A framework for proactive construction defect management using BIM, augmented reality and ontology-based data collection template. *Autom. Constr.*, 33: 61-71. https://doi.org/10.1016/j.autcon.2012.09.010

Pereira, F., Silva, C. and Alves, M. (2011). Virtual fitting room augmented reality techniques for e-commerce. *Enterprise Information Systems*, 220: 62-71. https://doi.org/10.1007/978-3-642-24355-4_7

Portman, M.E., Natapov, A. and Fisher-Gewirtzman, D. (2015). To go where no man has gone before: Virtual reality in architecture, landscape architecture and environmental planning. *Computers, Environment and Urban Systems*, (54): 376-384. doi: 10. 1016/j.compenvurbsys.2015.05.001

Protchenko, K., Dabrowsk, P. and Garbacz, A. (2018). Development and assessment of VR/AR solution for verification during the construction process. *MATEC Web of Conferences*, 196: 1-6. https://doi.org/10.1051/matecconf/201819604083

Puyan, A.Z., Wei, L., Dee, A., Pottinger, R. and Staub-French, S. (2019). BIM-CITYGML data integration for modern urban challenges. *Journal of Information Technology in Construction (ITcon)*, 24: 318-340. http://www.itcon.org/2019/17

Quinn, C., Shabestari, A.Z., Misic, T., Gilani, S., Litoiu, S. and McArthur, J.J. (2020). Building automation system – BIM integration using a linked data structure. *Automation in Construction*, 118: 103257. https://doi.org/10.1016/j.autcon.2020.103257

Rahimian, F.P., Seyedzadeh, S., Oliver, S., Rodriguez, S. and Dawood, N. (2020). On-demand monitoring of construction projects through a game-like hybrid application of BIM and machine learning. *Automation in Construction*, (110): 103012. doi: 10. 1016/j.autcon.2019.103012

Rankohi, S. and Waugh, L. (2013). Review and analysis of augmented reality literature for construction industry. *Vis. Eng.*, 1: 9. https://doi.org/10.1186/2213-7459-1-9

Ratajczak, J., Riedl, M. and Matt, D.T. (2019). BIM-based and AR application combined with location-based management system for the improvement of the construction performance. *Buildings*, 9(5): 118.

Razkenari, M., Bing, Q., Fenner, A. and Hakim, H. (2019). Industrialized construction: Emerging methods and technologies. *Computing in Civil Engineering, 2019.* ASCE International Conference on Computing in Civil Engineering, 2019. https://doi.org/10.1061/9780784482438.045

Research and Markets (2018). Augmented reality and virtual reality market by offering (hardware & software), device type (HMD, HUD, handheld device, gesture tracking), application (enterprise, consumer, commercial, healthcare, automotive), and geography-global forecast to 2023. Accessed February 7, 2019. https://www.researchandmarkets.com /research/qq837j/global_augmented?w=4

Rua, H. and Alvito, P. (2011). Living the past: 3D models, virtual reality and game engines as tools for supporting archaeology and the reconstruction of cultural heritage–The case study of the Roman villa of Casal de Freiria. *Journal of Archaeological Science*, 38(12): 3296-3308. https://doi.org/10.1016/j.jas.2011.07.015

Rymaszewska, A., Helo, P. and Gunasekaran, A. (2017). IoT-powered servitization of manufacturing – An exploratory case study. *International Journal of Production Economics*, 192: 92-105. doi: 10.1016/j.ijpe.2017.02.016

Sachs, R., Perlman, A. and Barak, R. (2013). Construction safety training using immersive virtual reality. *Construction Management Economics*, 31(9): 1005-1017. doi: 10.1080/ 01446193.2013.828844

Sachs, R., Treckmann, M. and Rozenfeld, O. (2009). Visualization of work flow to support lean construction. *Journal of Construction Engineering and Management*, 135(12): 1307-1315.

Sampaio, A.Z. and Martins O.P. (2014). The application of virtual reality technology in the construction of bridge: The cantilever and incremental launching methods. *Automation in Construction*, (37): 58-67. doi: 10.1016/j.autcon.2013.10.015

Schlueter, A. and Geyer, P. (2018). Linking BIM and design of experiments to balance architectural and technical design factors for energy performance. *Autom. Constr.*, 86: 33-43. https://doi.org/10.1016/j.autcon.2017.10.021

Shekhar, R., Dandekar, O., Bhat, V., Philip, M., Lei, P., Godinez, P., Sutton, E., George, I., Kavic, S. and Mezrich, R. (2010). Live augmented reality: A new visualization method for laparoscopic surgery using continuous volumetric computed tomography. *Surgical Endoscopy*, 24(8): 1976-1985.

Sheng, Y., Yapo, T.C., Young, C. and Cutler, B. (2011). A spatially augmented reality sketching interface for architectural daylighting design. *IEEE Transactions on Visualization and Computer Graphics*, 17(1): 38-50.

Shi, Y., Du, J., Ragan, E., Choi, K. and Ma, S. (2018). Social influence on construction safety behaviors: A multi-user virtual reality experiment. *Construction Research Congress*, 2018: 174-183. doi: 10.1061/9780784481288.018

Sidani, F.M., Dinis, L., Sanhudo, J., Duarte, J.S., Baptista, J.P., Martins, A. and Soeiro (2019). Recent tools and techniques of BIM-based virtual reality: A systematic review. *Arch. Comput. Methods Eng.*, 1-14. https:// doi. org/ 10. 1007/ s11831 - 019 - 09386 - 0

Simpfendörfer, T., Baumhauer, M., Müller, M., Gutt, C.N., Meinzer, H.P., Rassweiler, J.J., Guven, S. and Teber, D. (2011). Augmented reality visualization during laparoscopic radical prostatectomy. *Journal of Endourology*, 25(12): 1841-1845.

Siu, M.F.F., Lu, M. and AbouRizk, S. (2013). Combining photogrammetry and robotic total stations to obtain dimensional measurements of temporary facilities in construction field. *Visualization in Engineering*, 1(1): 1-15.

Stets, J.D., Sun, Y., Corning, W. and Greenwald, S. (2017). Visualization and labeling of point clouds in virtual reality. *In*: Proceedings of SIGGRAPH Asia 2017 Posters, Association for Computing Machinery, Bangkok, Thailand, pp. 1-2.

Su, X., Talmaki, S., Cai, H. and Kamat, V.R. (2013). Uncertainty-aware visualization and proximity monitoring in urban excavation: A geospatial augmented reality approach. *Visualization in Engineering*, 1(1): 1-13.

Tang, P., Akinci, B. and Huber, D. (2009). Quantification of edge loss of laser scanned data at spatial discontinuities. *Autom. Constr.*, 18(8): 1070-1083. https://doi.org/10.1016/j.autcon.2009.07.001

Tang, P., Anil, E.B., Akinci, B. and Huber, D. (2011). Efficient and effective quality assessment of as-is building information models and 3D laser-scanned data. *In*: Proceedings of the International Workshop on Computing in Civil Engineering, 2011, Miami, FL, USA, 19-22 June 2011; pp. 486-493. https://doi.org/10.1061/41182(416)60

Tao, F., Cheng, J., Qi, Q., Zhang, M., Zhang, H. and Sui, F. (2018). Digital Twin-driven product design, manufacturing and service with Big Data. *The International Journal of Advanced Manufacturing Technology*, 94(4): 3563-3576. doi: 10.1007/s00170-017-0233-1

Thiel, F., Discher, S., Richter, R. and Dollner, J. (2018). Interaction and locomotion techniques for the exploration of massive 3D point clouds in VR environments. *International Archives of the Photogrammetry, Remote Sensing and Spatial Information Sciences*, 42(4): 623-630. doi: 10.5194/isprs-archives-XLII-4-623-2018

Tromp, J.G., Le, D.N. and Le, C.V. (2020). *Emerging Extended Reality Technologies for Industry 4.0: Early Experiences with Conception, Design, Implementation, Evaluation and Deployment*. ISBN: 978-1-119-65463-6

Tsilimantou, E., Delegou, E.T., Nikitakos, I.A., Ioannidis, C. and Moropoulou, A. (2020). GIS and BIM as integrated digital environments for modeling and monitoring of historic buildings. *Appl. Sci.*, 10(3): 1078. https://doi.org/10.3390/app10031078

Vasilevski, N. and Birt, J. (2020). Analyzing construction student experiences of mobile mixed reality enhanced learning in virtual and augmented reality environments. *Research in Learning Technology*, 28. https://doi.org/10.25304/rlt.v28.2329

Vincke, S. and Vergauwen, M. (2019). Geo-registering consecutive datasets by means of a reference dataset, eliminating ground control point indication.

International Archives of the Photogrammetry, Remote Sensing and Spatial Information Sciences, XLII-5/W2, pp. 85-91. https://doi.org/10.5194/isprs-archives-XLII-5-W2-85-2019

Wagner, D. and Schmalstieg, D. (2003). First steps towards handheld augmented reality. *IEEE Xplore*, 127-135. doi: 10.1109/ISWC.2003.1241402

Waissi, G.R., Demir, M., Humble, J.E. and Lev, B. (2015). Automation of strategy using IDEF0 – A proof of concept. *Operations Research Perspectives*, 2: 106-113. https://doi. org/10.1016/j.orp.2015.05.001

Wakeman, I., Light, A., Robinson, J., Chalmers, D. and Basu, A. (2011). Deploying pervasive advertising in a farmers' market. *Pervasive Advertising*, Springer, London, pp. 247-267.

Wang, L.C. (2008). Enhancing construction quality inspection and management using RFID technology. *Autom. Constr.*, 17: 467-479. https://doi.org/10.1016/j.autcon.2007.08.005

Wang, X., Kim, M.J., Love, P.E.D. and Kang, S.C. (2013). Augmented reality in built environment: Classification and implications for future research. *Automation in Construction*, 32(1): 1-13.

Wang, X., Love, P.E. and Davis, P.R. (2012). BIM+AR: A framework of bringing BIM to construction site. *In*: Proceedings of the Construction Challenges in a Flat World, 2012, West Lafayette, IN, USA, 21-23 May 2012; Cai, H., Kandil, A., Hastak, M., Dunston, P.S. (Eds.). ASCE: Reston, VA, USA, pp. 1175-1181. https://doi.org/10.1061/9780784412329.118

Wang, X., Love, P.E.D., Kim, M.J., Park, C.S., Sing, C.P. and Hou, L. (2013). A conceptual framework for integrating building information modeling with augmented reality. *Autom. Constr.*, 34: 37-44. https://doi.org/10.1016/j.autcon.2012.10.012

Wang, X., Truijens, M., Hou, L., Wang, Y. and Zhou, Y. (2014). Integrating augmented reality with building information modeling: Onsite construction process controlling for liquefied natural gas industry. *Autom. Constr.*, 40: 96-105. https://doi.org/10.1016/j. autcon.2013.12.003

Wang, K.C., Wang, S.H., Kung, C.J., Weng, S.W. and Wang, W.C. (2018). Applying BIM and visualization techniques to support construction quality management for soil and water conservation construction projects. *In*: Proceedings of the 35th International Symposium on Automation and Robotics in Construction (ISARC 2018), Berlin, Germany, 20-25 July 2018, volume 35. https://doi. org/10.22260/ISARC2018/0099

Webster, A., Feiner, S., MacIntyre, B., Massie, W. and Krueger, T. (1996). Augmented reality in architectural construction, inspection and renovation. *In*: Proceedings of ASCE Third Congress on Computing in Civil Engineering, Vol. 1, p. 996.

Wolfartsberger, J. (2019). Analyzing the potential of virtual reality for engineering design review. *Automation in Construction*, 104: 27-37. https://doi. org/10.1016/j.autcon.2019.03.018

Woodward, C. and Hakkarainen, M. (2011). Mobile mixed reality system for architectural and construction site visualization. *Augmented Reality – Some Emerging Application Areas*, Intechopen, 115-130.

Wu, W., Hartless, J., Tesei, A. and Gunji, V. (2019). Design Assessment in virtual and mixed reality environments: Comparison of novices and experts. *Journal of Construction Engineering and Management*, 145(9). https://doi.org/10.1061/(ASCE)CO.1943-7862.0001683

Wu, I. and Hseih, S.H. (2012). A framework for facilitating multi-dimensional information integration, management and visualization in engineering projects. *Automation in Construction*, 23: 71-86.

Yan, W., Culp, C. and Graf, R. (2011). Integrating BIM and gaming for real-time interactive architectural visualization. *Automation in Construction*, 20(4): 446-458.

Zhang, J., Teizer, J.K., Lee, C.M., Eastman and M. Venugopal (2013). Building Information Modeling (BIM) and safety: Automatic safety checking of construction models and schedules. *Automation in Construction*, 29: 183-195. https://doi.org/10.1016/j.autcon.2012.05.006

Zhang, D., Zhang, J., Xiong, H., Cui, Z. and Lu, D. (2019). Taking advantage of collective intelligence and BIM-based virtual reality in fire safety inspection for commercial and public buildings. *Appl. Sci.*, 9: 5068.

Zhao, D. and Lucas, J. (2015). Virtual reality simulation for construction safety promotion. *International Journal Injury Control Safety Promotion*, 22(1): 57-67. doi: 10.1080/ 17457300.2013.861853

Zhou, Y., Luo, H. and Yang, Y. (2017). Implementation of augmented reality for segment displacement inspection during tunneling construction. *Automation in Construction*, 82: 112- 121. https://doi.org/10.1016/j.autcon.2017.02.007

Zhu, W., Owen, C.B., Li, H. and Lee, J.H. (2004). Personalized in-store e-commerce with the promopad: An augmented reality shopping assistant. *Electronic Journal for E-commerce Tools and Applications*, 1(3): 1-19.

Zollmann, S., Hoppe, C., Kluckner, S., Poglitsch, C., Bischof, H. and Reitmayr, G. (2014). Augmented reality for construction site monitoring and documentation. Proceedings of the IEEE, 102(2): 137-154. doi: 10.1109/JPROC.2013.2294314

Smart Maintenance Services for Buildings with Digital Twins and Augmented Reality

**Stephan Embers, Patrick Herbers, Markus König*, Mario Wolf
and Sven Zentgraf**

Chair of Computing in Engineering, Faculty of Civil and Environmental
Engineering, Ruhr-Universität Bochum, 44801 Bochum, Germany

6.1 Introduction

The product-use phase is the most cost-intensive phase in the product lifecycle of industrial goods and buildings. The direct annual expenditure for maintenance amounts to about 250 billion Euros and thus to about 10 per cent of the German Gross Domestic Product (GDP) (Kuhn and Bandow, 2009). Maintenance measures are carried out to extend the lifespan of capital goods. Depending on the frequency of maintenance, the cost of upkeep can vary. If maintenance work is delayed or not carried out at all, important systems or parts of a building can break down much sooner than their life expectancy would suggest.. This then must be replaced at greater expense. On the other hand, if the frequency of maintenance is too high, the cost of labor might exceed the benefits of keeping old equipment in working condition. In order to minimize the cost associated with the maintenance frequency, an optimal balance needs to be found for the aforementioned cases (Lee and Scott, 2009). To help find and establish this balance, a smart maintenance system can be deployed (Bokrantz *et al.*, 2020). Smart maintenance uses various technologies, such as sensors to monitor a building and to quickly identify possible problems and alert maintenance staff for repairs (Pašek and Sojková, 2018).

_____

*Corresponding author: koenig@inf.bi.rub.de

For a smart maintenance system to correctly and accurately predict when and where maintenance is required, it needs sensor data of a building. The sensor data is gathered by sensors placed within the building. These sensors are then virtually connected with a Digital Twin to localize them. With the help of the gathered sensor data, the smart maintenance system is then able to check for errors and malfunctions within the building, using its Digital Twin. It also knows what is causing the problem and where it can be found. Furthermore, it can calculate the best course of action to repair a malfunction and instruct maintenance personnel how to repair it and suggest what is needed for repairs (Boschert *et al.*, 2018).

To report an error or malfunction, a system first needs to know if there is a problem to report. For this, the system must know the behavior of an object or system and if it is working correctly. To help the system identify an object and its function, it must know its semantics. The semantics of an object show the system what an object should do and how it should function so that the system can infer when an object is working improperly. Knowing the location of an object can help the system with the semantics. It also helps in identifying the specific object. Also, knowing the location of an object can guide maintenance personnel directly to the object in case of malfunction. With the help of an object's semantics and location, the system can check the object using current sensor data to detect a possible malfunction and call maintenance personnel to repair it if need be.

In this chapter, we will look at the different ways that could help to implement such a smart maintenance system. We will introduce the BIM-methodology and the concept of a Digital Twin. Further, the methods of establishing a link to a Digital Twin of the building as well as identifying and localizing an object will be presented. Additionally, case studies will be shown on how these technologies have already been used in maintenance.

6.2 Building Information Models

The planning and realization of cognitive buildings is even more elaborate and complex than traditional building projects of similar size. Many trades need to coordinate their planning, but especially in the field of cognitive buildings this is almost impossible to achieve using analog methods, such as 2D construction plans and drawings. The inspection of plans is usually done manually, resulting in highly error-prone steps carried out by the planners of the involved disciplines. This analog procedure can lead to time-consuming and cost-intensive errors, especially when changes are made during both the planning and construction process.

In order to overcome these limitations and to generate a more efficient way of planning, building and operating structural systems, the method of Building Information Modeling (BIM) has been developed over the last decades and is still being continuously expanded today. At the center of this methodology is the Building Information Model. It consists of a three-dimensional (3D) model of the building project to be realized. During the lifecycle of the building, all relevant information and data of the stakeholders are linked to the 3D model (Fig. 6.1). From this, various plans, simulations and as-is states can be generated (Borrmann *et al.*, 2018b).

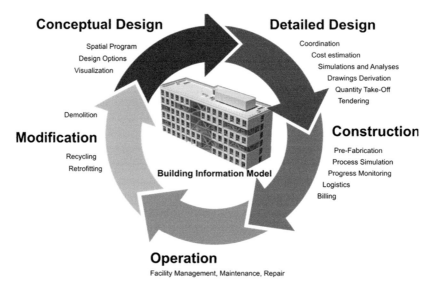

Fig. 6.1: Illustration of the BIM lifecycle (Borrmann *et al.*, 2018a)

At the beginning of the implementation, namely the conceptual planning phase, the Building Information Model mainly contains, in addition to the basic building model, further design variants for the planned building, including visualizations of the building. Also, in this process, specifications for rooms and dedicated spatial areas of the planned volume can be generated and bound to the model. These can be simple specifications for room sizes or more complex plans, such as schemes for room usage in each floor or specifications for the desired lighting concept.

After completion of the conceptual planning, models of the involved trades can be linked in a more detailed design process. These models include plans for HVAC, fire protection, lighting, and many others. This has the advantage that simulations and analyses can be used to detect collisions caused by decisions made by various trades at this early stage of the project, which would cause additional workload and costs at later stages. At the end of the planning process, 2D plans can be derived from the extended model, which are collision free. Furthermore, initial quantity estimations of the required materials and components can be calculated. In addition, all information necessary for tendering of the project can be generated from the linked data of the individual trade planners.

Before the transition to the actual construction phase, all parts can be extracted from the Building Information Model that can be produced in prefabrication processes off-site. A schedule can be used in combination with the 3D model to create a process-flow simulation to plan orders for needed materials and personnel more efficiently. Such a linked model is also called a 4D model in the BIM methodology, since time is now included as an additional dimension. In addition, this 4D model can be used to carry out renewed clash detection in order to check the planned construction processes for errors before the actual construction begins. Parallel to material and personnel planning, such a 4D model can also be used for process monitoring, which shows which construction segment is in which production stage at which time. A third point that can be handled using such a 4D model is the planning of safety-relevant equipment, such as fall protection systems. A so-called 5D model, now extended with a 'cost' dimension, contains all details about the costs of the materials and components required. This makes it possible to calculate costs according to the construction process and to generate invoices during the construction process. In accordance with an optimal BIM process, all data concerning the materials and components used are entered into the model during the implementation in order to create an even more detailed digital twin of the building project (Borrmann *et al.*, 2018a; Zapata *et al.*, 2019).

6.3 Digital Twin

The term 'Digital Twin' is used to describe the virtual representation of a real object. The Digital Twin not only reflects the geometric representation of a building, but also includes further information, such as the material of the components, simulations of building behavior, and many other relevant data. It should be noted that the physical counterpart of the Digital Twin need not necessarily exist. Especially in production engineering or even in civil engineering, as described in the previous chapter, such a twin is used to realize the planned construction. A BIM model can thus be considered as the Digital Twin of a building during planning, construction, and operation (May, 2018).

In the field of mechanical and plant engineering, Digital Twins have long been consulted for the planning, construction, and operation of machines and plants. With first drafts, simulations can already be carried out in the early stages of realization in order to later obtain a product that is as error-free and easy to use as far as possible. With the help of the twin, construction and commissioning can be supported in the form of videos and other types of instructions, for example, to be used within AR frameworks. In addition, Digital Twins can be used to make diagnoses during operation and to read data from built-in sensors in order to have exact information about the state of the real counterpart at any time. This method is called Digital Twin linking (Fig. 6.2).

Fig. 6.2: Faculty building of the Bochum College (*left*) and its Digital Twin (*right*)

Compared to the industries just presented, the use of Digital Twins in the construction industry is less widely used. This is mainly due to the fact that, as described in the previous chapter, the BIM methodology provides a guideline for the treatment of such a twin, but due to limitations it cannot be fully implemented (Grosse, 2019). Usually, the building model is handed over to the building operator after acceptance of the construction project. The data created and linked during the construction phase can then be transferred directly from the building operator to a facility management tool. To make use of the Digital Twin, the listed information on materials and components are used to support maintenance processes during the operation time of the building. In the field of cognitive buildings, even regular maintenance processes could be transferred from the model to the responsibility of the cognitive system. The cognitive building could then instruct maintenance technicians in detail as to where maintenance is required and what is needed to fix a problem.

It is important for all maintenance and renovation processes to be continuously transferred to the model in order to keep the Digital Twin of the building up to date. This is particularly relevant for major renovations and also for the demolition of the building, in order to be able to plan how much construction waste will be produced, which hazardous substances will have to be removed, and which materials and components can be recycled in the best case (Borrmann *et al.*, 2018a).

The main problem is that, through the involvement of various trades from different disciplines each using their own software tools and data formats, the components are often not compatible with each other, making it difficult to link the necessary data to form a Digital Twin. In order to counteract this problem, organizations, such as buildingSMART Deutschland e. V. (https://www.buildingsmart.de/impressum) are working on an open data standard, such as the Industry Foundation Classes (IFC) which should create an open exchange format for all trades. In Germany and Europe particularly there are many government-funded research projects that support this process. In this context, solutions are also being sought to optimize and simplify some of the processes that can currently be carried out manually only by using rule-based test systems and digital systems instead. However, before a solution for the all-encompassing use of Digital Twins in construction can be developed, especially in the field of cognitive buildings, many research steps still need to be implemented (Borrmann *et al.*, 2018c).

In particular, AI, IOT, and IIOT need to be developed effectively as they play a key role in the dissemination and usability of Digital Twins. For a deeper insight into the topic of Digital Twin and the combination of the above-mentioned technologies, please refer to the work of Fuller *et al.* (2020).

6.4 Maintenance

Maintenance is defined by the German Institute for Standardization in (DIN 31051, 2019) as a combination of all measures applied during the lifecycle of a unit that serves to maintain or restore its functional condition. The field of maintenance is divided into four subcategories: maintenance, inspections, repairs, and improvements.

Maintenance itself is carried out to counteract the reduction of the wear and tear on the stock. According to the definitions in (DIN 13306, 2018), this refers to the harmful change of the physical condition due to time factors, use or external causes. Inspections are used to determine and evaluate the actual condition of a unit, including determining the causes of wear and tear and deriving the necessary consequences for future use. Repairs are physical measures that are performed to restore the function of a defective unit. In contrast, improvements serve to increase the reliability and/or maintainability and/or safety of a unit without changing its original function. Improvements are achieved by technical, administrative, or management measures. Typical Key Performance Indicators (KPI) for the assessment of maintenance activities carried out are defined in (DIN EN 15341, 2007).

While maintenance is defined as a combination of all the measures during the lifecycle of a unit that serves to maintain or restore its functional condition, the functionality of a technical unit is only the basis for further objectives that are pursued with maintenance (Strunz, 2012). Figure 6.3 shows the subdivision of maintenance objectives according to Strunz (2012) into the optimization fields of cost objectives, safety objectives, and production objectives. Economic goals specially, such as the optimization

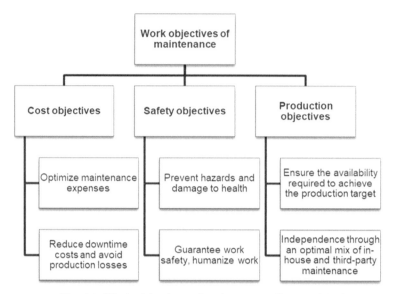

Fig. 6.3: Work objectives of maintenance (Strunz, 2012)

of maintenance expenditures, reduction of downtime costs, achievement of high availability, etc. require a maintenance strategy that does not only react to failures (Weidemann and Wolff, 2013).

Cost targets are the easiest to capture economically, since every maintenance activity involves direct costs, such as labor and material costs, but three to five times higher costs are incurred through indirect costs, such as a loss of production during machine downtime (Kuhn *et al.*, 2006). Safety objectives are difficult to record due to a wide variety of influencing factors, since accidents that have been successfully avoided are rarely reported, even though special systems exist for this purpose (Gnoni and Saleh, 2017). In addition to the increase in efficiency of the actual maintenance processes and the advantages of consistent data management, the occupational safety aspect is also improved by technical support (Acatech, 2015). Production goals in maintenance can be achieved by ensuring reliability. This basic idea is reflected in the changed business models in Industry 4.0, such as availability guarantees, the increased demands on reliability, and the ability to produce Just In Time (JIT) (Pawelek, 2013).

Furthermore, different maintenance measures can be differentiated according to their qualification requirements presented in (DIN 13306, 2018), which are summarized in Table 6.1. This is particularly relevant for the design of multimedia assistance systems, since these require particularly high success rates for routine work that is rarely performed (Schlick *et al.*, 2014). Detailed requirements for the qualification of maintenance personnel are listed in (DIN EN 15628, 2014), whereby the roles of maintenance specialists, maintenance managers, and maintenance engineers are distinguished.

Table 6.1: Levels of Maintenance Tasks as per Qualification
According to (DIN 13306, 2018)

Level	Description
Level 1	Simple measures, which are carried out after minimal training.
Level 2	Basic actions, which are carried out by qualified personnel according to detailed procedures.
Level 3	Complex measures, which are carried out by qualified technical personnel according to detailed procedures.
Level 4	Actions involving knowledge of a particular technique or technology and carried out by technical staff specializing in that technique or technology.
Level 5	Measures that include specialised knowledge of the manufacturer or a specialist company that has industrial supply or support equipment.

The standard updated 2014 lists as one of the required minimum skills for maintenance specialists, the application of technological tools, and the assurance that they are properly used.

DIN 13306 divides the maintenance strategies for achieving the various objectives into corrective and preventive maintenance. Additionally, Mobley (2002) describes the concept of predictive maintenance. Corrective maintenance, also known as 'fix it when it breaks', is the most basic form of maintenance, and demands almost no preplanning. For example, when an object requires maintenance, it is reported by the staff or residents, and a crew is sent out to repair it. It is by nature unscheduled and therefore can cause significant downtime of equipment. To mitigate the need for corrective maintenance, preventive maintenance calls for regular checkup of equipment and structural elements. Early signs of deterioration can be identified, and equipment can be replaced before it breaks, reducing the aforementioned downtime. Structural elements of buildings especially require preventive maintenance, as structural failure can cause major disasters and loss of human life. Checkups are usually scheduled monthly, annually, or per decade, depending on the domain.

Predictive maintenance is a concept that aims to optimize this schedule. Although corrective and preventive maintenance are still the most commonly employed approaches to maintenance today, there is a considerable demand for predictive systems in the industry (Werner *et al.*, 2019). Using data from past maintenance work, the time, a specific piece of equipment runs until failure, can help predict when the next preventive step needs to be taken. This data can, among others, consist of the previous failure rate, weather conditions, humidity, usage statistics, and power consumption. Enhancing preventive maintenance measures with predictive ones can improve efficiency and effectiveness of maintenance work (Lughofer and Sayed-Mouchaweh, 2019). However, predictive maintenance should not be seen as a substitute to preventive and corrective maintenance, as unexpected breakdowns are still possible even in the best maintained buildings (Mobley, 2002).

6.4.1 Maintenance in Cognitive Buildings

How can we utilize aspects of cognitive buildings to reduce the amount of corrective maintenance, introduce more predictive maintenance concepts, and make the actual maintenance work easier and more efficient? If we look at the maintenance routine itself, some areas can be identified as easily enhanceable. During a corrective maintenance procedure, the problem first needs to be reported and identified. In cognitive buildings, these problems can already be automatically reported by sensors in the building or by the broken device itself, without any need for human interaction. Preventive maintenance can be scheduled optimally in a cognitive building, reducing downtime. Once a problem has been identified or a scheduled maintenance task comes up, the service worker at hand requires information about the task: a description and the location of the necessary equipment as well as a list of materials, tools, permits, and safety equipment required for the task. For scheduled preventive maintenance, all this data should already be available, possibly in a digital format through a maintenance management system. Modern machine learning systems can benefit from the large amount of data to make informed decisions (Susto *et al.*, 2015). For corrective maintenance, this information needs to be gathered first. A cognitive building could, for example, automatically order replacement parts, acquire the necessary security permits, inform the residents, and calculate the optimal route for the worker through the building. Once a worker has arrived at the scene, she requires manuals and blueprints to access the location and identify the problem. Augmented reality can help display the required information and can give step-by-step guide for standard procedures. If the worker has finished the task, she can give detailed information back to the system to improve the predictive maintenance data. If the task requires more work or specific replacement parts, additional resources can be requested immediately, or the task can be rescheduled.

At this point we would like to remind the reader that it may be tempting to automate everything in the maintenance pipeline with minimal human interaction, but caution is advised. Maintenance is inherently a field that deals with uncertainty and unplannable consequences.

Automated systems and machine learning algorithms are susceptible to misinterpreting unknown situations. An overreliance on algorithms, instead of human reasoning, will cause harmless situations to be blown out of proportion, while serious and infrequently occurring problems are missed. A human element is necessary and welcomed, especially in the field of maintenance. The worker should still have the authority over the system, not the other way around, as the system is a tool instead of an overseer. Maintenance structures in cognitive buildings should therefore factor in the human as an integral and decisive element.

To develop a cognitive maintenance procedure, one needs to examine the tasks for which a system needs to be built. Inspection tasks are the most rudimentary of maintenance tasks. The focus of inspection is mainly on data acquisition, which can be gathered by humans as well as automated sensors. In a cognitive building, sensory information and human inspection routines feed into a pool of data, which can further be analyzed for predictive maintenance systems (Chen *et al.*, 2019). Inspection tasks can also be improved through augmented reality displays or point cloud captures. Combined with a form of indoor localization, inspection data can be compiled and analyzed in a central location. Other types of tasks, such as calibration tasks, also benefit from centralized data gathered through inspections and sensor readings. Calibration of utilities or HVAC systems can be done almost completely automated and (in a cognitive city) in accord with the city-wide grid.

In the section titled Case Studies, we will show current research in cognitive maintenance with the aid of augmented reality and Digital Twins in the field of mechanical engineering.

6.5 Mixed Reality

While there is no single definition for the term 'mixed reality' (Speicher *et al.*, 2019), it commonly refers to a set of visualization methods that combine reality and virtual information in different ways. Mixed reality thus offers the ideal interface for context-sensitive visualization. Milgram *et al.* (1994) describe the range of Mixed Reality characteristics as the reality-virtuality-continuum. The continuum (Fig. 6.4) exists between the cornerstones of actual reality and virtual reality. Between these extremes, there are two further areas called augmented reality and augmented virtuality.

Starting from actual reality, the density of virtual information or content increases steadily until no real information is displayed in virtual reality. The terms 'augmented reality' and 'augmented virtuality' are used to describe the introduction of either real or virtual information in the respective opposite representation. In this context, augmented reality is defined as an overlapping of reality with virtual content (Azuma, 1997).

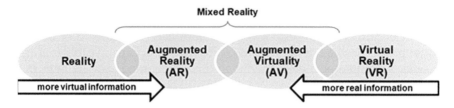

Fig. 6.4: The reality-virtuality-continuum according to (Milgram *et al.*, 1994)

This virtual content can take any form, for example, as text, images, or videos up to complex, interactive 3D models. The transition to augmented virtuality is fluent. On the AR side, augmented virtuality refers to the integration of real objects in a virtual environment and is reduced in the direction of virtual reality to the use of, for example, real operating data for the animation of virtual models. For the use of different manifestations of mixed reality, there are nevertheless well distinguishable implementation options. Rekitimo and Nagao (1995) have established some principles which are extended in Fig. 6.5. The typical use of graphical user interfaces (GUI, No. 1) on monitors is based on a clear separation of the real world and virtual content (Fig. 6.5).

In augmented reality (No. 2), humans can still manipulate the real world, but they can also view the virtual content displayed as an overlay over reality. Depending on the implementation, interactions of the environment with the AR application are also possible. In virtual reality (No. 3), the user can interact with the virtual content but is completely isolated from the real environment. In addition, the virtual content is also unrelated to the environment. The greatest variance here is shown by augmented virtuality (No. 4). Here, an immersion can exist, that is, the user immerses into a virtual world that surrounds her and contains elements or data of the real world, or there is no immersive representation so that, as in augmented reality, 3D representations for the visualization of real measured values are displayed without an overlap of virtual contents with reality.

6.5.1 Augmented Reality

The most frequently cited source in the field of augmented reality is the definition of AR according to Azuma, according to which AR is the combination of reality and virtuality through superimposition, which is displayed in real time, is interactive, and in which 3D objects can be used (Azuma, 1997). The representation of such an augmented reality can be divided into five steps: video recording, tracking, registration, display, and output (*see* Fig. 6.6), whereby these steps are performed per camera image or AR overlay frame (single image).

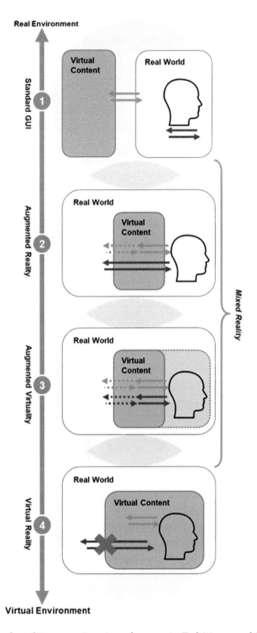

Fig. 6.5: Mixed reality overview in reference to Rekitimo and Nagao (1995)

The video recording is initially made by any camera whose parameters are known (webcam, camcorder, smart device camera, etc.). Tracking refers to the calculation of the position and orientation of the

Repeat every frame

Fig. 6.6: The five steps to calculate an AR view (Dörner *et al.*, 2013)

field of view of the used camera. Depending on the characteristics of the AR, this field of view differs from that of the AR user. The position can be captured with sufficient accuracy using inertial sensors, gyro sensors, and magnetometers. For position detection, a distinction must be made between indoor and outdoor areas and the required accuracy. For low-accuracy requirements in outdoor areas GNSS (Global Navigation Satellite System), or specifically GPS (Global Positioning System) or Galileo can be used. For indoor applications and (potentially) higher accuracy, computer vision approaches are used, which can record not only the absolute position of the camera, but also the position relative to a reference and are particularly suitable for mobile applications due to their high performance and their readily available cameras (Tönnis, 2010; Billinghurst *et al.*, 2015).

For tracking, a distinction is made in the field of augmented reality between the two most widespread methods of camera-based tracking, namely using markers or doing feature-based tracking [33]. Another tracking approach that works without requiring prior data generation or a known reference in the environment is Simultaneous Localization and Mapping (SLAM). This approach originates from robotics and gradually creates a map of the environment by feature extraction and by tracking of device movement. Many modern AR SDKs support both marker-based tracking and SLAM tracking (Koch *et al.*, 2014), so they can be used in combination. Thus, an initial positioning via a marker is detected and then switched to SLAM tracking. The advantage of combining these methods is the secure initialization, after which tracking and orientation can be done without tracking markers having to remain in view.

Registration also means anchoring the digital data in the real world. The simplest case here is the registration of a 3D object in the origin of the coordinate system of a tracking marker. The points display and output depend on the technical implementation or the output device.

6.5.1.1 Optical-See-Through (OST)

In the optical see-through procedure, the previously mentioned steps of video recording, tracking, and registration are carried out and then a perspective correction of the 3D content based on the camera pose is calculated (Dörner *et al.*, 2013; Tönnis, 2010). By using, for example, semi-transparent mirrors, the user can perceive the environment without hindrance, even in the event of a system failure. One problem with using semi-transparent mirrors, however, is that they can glare in bright environments, which can blur the overlay. Azuma (2016) sees the challenge for the technical development of the next HMD generation in the local shading of mirrors to counteract this effect. Only the previously perspective-corrected 3D content is output on the display and 'mixed' with reality by reflection. However, due to the fact that the mirror is mounted just before the eyes, there is a difference in the focus in the field of vision of the viewer (Van Krevelen and R. Poelman, 2010). The field of view of the AR glasses available on the market varies considerably, which is why 'cut-off' effects can also occur with the latest generation of high-priced devices, where the overlay ends well in front of the natural field of view, as explained in Tönnis (2010).

6.5.1.2 Video-See-Through (VST)

With the video-see-through procedure, the same steps are carried out as before (video recording, tracking, registration, display), only that now both the camera image itself and the display of the 3D content must be calculated before the output is made (Dörner *et al.*, 2013; Tönnis, 2010). The user of the HMD is looking at a display which is mounted just before her eyes. Fresnel lenses are often used to compensate for the very short distance between the eyes to prevent irritation (Dörner *et al.*, 2013). Since the user perceives both the environment and the digital overlay via the display, a system failure leads to a 'blackout'. The user must first remove the HMD in order to see the reality with her own eyes. A clear advantage of the system is the better quality of the overlay, as the shading of the overlay only depends on the camera image and not on an optical element or the environment (Tönnis, 2010). The same principle is applied to handheld AR when smart devices are used to display the AR content, where all essential elements are covered by the sensor technology of the smart device. The disadvantages of 'blackout' are eliminated by the normal working distance to the smart device, but the smart device must be held in the hand, thus limiting the ability of technicians to work hands-free (Van Krevelen and R. Poelman, 2010). However, some advantages of smart devices are their performance, connectivity, low cost, and usability as a general working tool.

6.5.2 Mixed Reality in Maintenance

Each form of mixed reality has specific use cases, strengths and weaknesses, which also depend on the state of the art and the respective device. In this context, current economic figures show that virtual reality technology has already arrived on the consumer market with over one million devices sold by the end of 2017, while augmented reality is largely perceived as still in the development phase (Lamkin, 2017). These numbers have changed significantly since then, with 5.7 million VR devices sold in 2019 alone (SuperData, 2019).

The applications in the field of maintenance can be distinguished by the type of mixed reality. It quickly becomes clear that the assistance systems for on-site maintenance are usually implemented using augmented reality, whereas virtual reality and augmented virtuality are used for training purposes. Dini and Mura showed in (Dini and Mura, 2015) that 54 per cent of all activities supported by augmented reality in a lifecycle spanning environment are performed in maintenance and another 24 per cent in inspections. Furthermore, Dini and Mura showed in the same paper that the majority (78 per cent) of AR applications are implemented by video see-through procedures and that 58 per cent of research uses HMDs and 31 per cent use handsets.

The intuitive presentation of augmented reality leads to an improvement in error rates, especially in activities that are not performed regularly (Schlick, 2014). Further studies show that context-sensitive maintenance support with Augmented Reality reduces working time (Hou *et al.*, 2013; Lamberti *et al.*, 2014) and error rates (Hou *et al.*, 2013; Abramovici *et al.*, 2014) compared to classical documentation types.

6.6 Object Identification

A fundamental problem of technology that interacts with the environment is understanding our surroundings. While a human has no problem identifying what an object is, what it is used for, and how it works, a machine does not have the advantages of a human brain. Since general intelligence has not yet been solved, we require some form of heuristic to help a machine extract knowledge from its sensory information.

A key differentiation in object identification is semantics and identity. Semantics is knowledge about what an object is, what it can do, and what it is used for. Identity is recognizing an object as that specific object. While semantics describes what an object is, it does not include knowing the identity of the object. A system that is able to attribute semantic knowledge to an object might later be able to attribute the same semantics to that object, but cannot be certain of its identity. For example, a neural network that is able to recognize cats will be able to recognize the same creature as a cat at different times, but it cannot say that it is still the same cat. Likewise, a system that can determine the identity of an object does not require semantic knowledge. A Quick Response Code (QR Code) scanner does not know about the semantics of the code it is scanning, but it is able to recognize the previously seen codes.

Semantics and identity are not completely separate, though; one can help to infer the other. If the semantics of an object are known to be unique, it is congruent with the identity. If the semantics of an object is already known, for example by previously storing it in a database, the identity can be used to look up semantic information about the object. Other factors, such as location or time, can also help in determining semantics or identity. It is important when choosing a method for object identification to be aware what kind of information is needed, and in which priority. Several approaches exist for both identity and semantic extraction, with different advantages and disadvantages depending on the task, the complexity, dependencies on outside sources, deployment effort, etc.

6.6.1 Markers

The simplest and most straightforward way of determining identity is by simply labeling the object in question with some form of identifying marker. This can range from a common serial ID on an object to integrated RFID markers. Visual markers usually appear in machine readable form: barcodes, QR codes or AR-markers. Visual markers have a low initial cost, require virtually no maintenance, and can be read by simple visual sensors such as common smartphone cameras. To extract semantic information, an ID in a visual marker can be queried from a database which stores previously-defined semantic information about an object. Barcodes are the most common form of object identification, being easy to read and visually unobtrusive. A disadvantage of barcodes is the limited information they can store, and they provide no alignment features. For augmented reality, proper digital overlays over an object (sometimes called holograms) are only possible if the object can be visually aligned from the camera input, which requires determining the relative orientation and position (pose) of the viewing camera in relation to the object. Two-dimensional barcodes, such as QR codes, have the advantage of conveying orientation and perspective information for the camera pose estimation. QR codes consist of three alignment points which allow the computation of viewing angles through a reverse perspective transform. Additionally, QR codes can store information more densely than barcodes. Additional information about the object could thus be stored offline without access to a database. AR markers are similar to QR codes, as both are two-dimensional markers.

Fig. 6.7: A barcode, a QR code, and a common AR marker

Instead of being designed to store information, AR markers are easily recognized and rectified at large distances and extreme angles (Hubner *et al.*, 2018). Of course, visual markers also have disadvantages. Depending on the application and the device used for interaction, visual markers may prove insufficient for certain tasks. Device drift may become a problem when the visual marker is not in sight – a problem that can occur when inspecting larger objects. Line of sight must be reestablished after a short time to correct inaccuracies in pose. One solution would be to add more markers, but even in a cognitive building one must regard aesthetics. Invisible electronic markers, such as RFID, have the advantage of being unintrusive, but are connected to higher costs, short distances, and do not allow for accurate pose estimation (Sanpechuda and Kovavisaruch, 2008). Another aspect is that markers only work on objects that have previously been defined in a database or adhere to a common standard of marking. Unknown objects without markers or objects with unknown markers cannot be identified by this technique.

6.6.2 Image Recognition

For semantic extraction, image recognition is one of the most popular methods. Objects are connected to a semantic understanding by recognizing shape and color and assigning a previously learned category. Image recognition also allows for marker independent-pose estimation, given that the object in question is made up of sufficiently distinguishing features. Recognition methods are best suited for detecting unmarked building elements, such as windows, doors, or non-stationary equipment. Image recognition is also commonly used for natural marker detection. Natural markers, such as signage, require no additional cost as they are already present in almost any building. While offering similar advantages to normal markers, they require more sophisticated methods for detection than markers specifically designed to be machine readable. Usually, image recognition is done by using machine learning techniques, such as Convolutional Neural Networks (CNNs). CNNs are able to locate and classify a wide range of objects from image feeds, which would allow for semantic association without markers. For image recognition, a building would not require any additional changes. Unfortunately, neural networks require an exhaustive training phase and are non-trivial to design. Training requires large amounts of pre-labeled data to be effective and cannot recognize objects that are not part of the training data. Once trained, CNNs are fast to execute and well-designed networks can reach high accuracies in detection and classification. Additional sensors may improve detection and tracking, such as depth sensors or inertial measurement units. Image recognition is best used when a semantic understanding of commonly occurring and visually distinct objects is required.

6.6.3 Digital Twin Linking

The concept of Digital Twin linking is central to augmented maintenance in cognitive buildings. Digital Twins are virtual projections of real-life objects that can hold semantics or data, or offer functionality. Common Digital Twins are digital building models, which correspond to a specific building. Modern buildings are commonly designed digitally, with BIM models readily available. Maintenance tasks can directly benefit from having access to the BIM model, even more so if a direct link exists on an AR device to read and manipulate data, or operate facility functions. Other types of Digital Twins are used for stationary and non-stationary equipment. Assembly and maintenance instructions can be displayed on augmented reality devices in the form of a Digital Twin. Displaying data or instructions where they are needed can help alleviate the cognitive load on a worker (Baumeister, 2017), improving safety and efficiency. Linking a Digital Twin to a real object requires either knowing the identity or the semantics of an object, depending on the task. Note that one can infer the semantics of an object based on its identity if a Digital Twin is present, but not the other way around. Thus, the identity is preferable, but not required for all the tasks; for example, device manuals in augmented reality only require knowledge of the semantics of an object, not the identity. The linking can be realized on three different levels: data linking, functional linking, and pose linking.

Data linking allows fetching data from or writing data to a Digital Twin. This can be realized as a manual or automated process. Displaying technical data or adding notes to a specific object are typical uses of data linking. Data linking does not require an external source of information, such as a database, but is often done so.

Functional linking allows direct control of an object and is inherently a user-invoked process. Examples of this are virtual elevator buttons or remotely controlled valves. Functional linking thus always requires some form of communication between a device and the object over which to send commands. Functional linking can be seen as an extension of the Internet of Things (IoT), in which devices can be accessed and controlled through a network connection. Similar to IoT, security becomes a concern for a cognitive building. Systems must be designed with great care to avoid security breaches.

Pose linking is an important aspect for augmented reality applications. If the poses of the object, the viewport, and the Digital Twin can be mapped, the real object and the virtual object become one single augmented object. Virtual information can be displayed directly on the real object. Pose linking Digital Twins of buildings, also referred to as indoor localization, enables tasks that rely on user location, such as indoor navigation. While pose linking is essential for immersive AR, it is also the most difficult type of linking to achieve. If the AR device is not static, which is usually the case, an initial linking process is required. In case of linking the Digital Twin of a building, the device has to guess where, in the building, it currently is located. This can be done through markers or image recognition tools, or with the help of the user. After linking the position and rotation of the device for the first time, it needs to be constantly tracked to update the pose. Various sensory techniques exist and are already built into modern AR devices, but the pose may still drift after long usage. At the moment, marker-based pose linking is the most reliable type. Machine learning could in the future improve markerless pose linking to a point where no markers would be necessary in a cognitive building.

All three types of linking can be combined: functional linking often requires data linking to provide information about the current state of the object. Pose linking can support both data linking and functional linking by showing contextual information or controls through holograms. When designing a cognitive building, one has to keep in mind what kind of linking processes are needed, and which techniques for determining the identity or semantics should be supported.

6.6.4 Indoor Localization

A special case of pose linking is indoor localization. When linking the pose of a device to a digital building model, it is equivalent to locating the device inside the building. An accurate localization would allow for location-dependent tasks, such as indoor navigation, which is still an unsolved problem. While outdoor localization is usually done through GPS, the unreliability that is prevalent inside of buildings forces us to utilize other sensors to obtain an accurate pose indoors.

The three most common applications for indoor localization are indoor navigation, device tracking, and pose linking of static objects. Indoor navigation allows a user to find her way in an unknown building and use the shortest path in large structures. It can also offload finding a shelf in a warehouse or a specific room in an office. Indoor navigation may become as prevalent in modern buildings as GPS navigation is on the road. On a similar note, knowing the location of a device also allows tracking it. Especially in medical fields, knowing the location of equipment or staff can be crucial. Lastly, indoor localization can be seen as a method of pose linking of static objects inside a building. While indoor localization can be seen as equivalent to pose linking, localization does not require the full pose depending on the task. Only knowing the general location can already enable data linking to the room a user is currently occupying, or offer basic navigation and tracking.

The quality of localization depends on a few factors, mainly accuracy, reliability, performance, and cost. Accuracy can be described as the average distance between the Digital Twin and the actual model after localization. Pose linking for augmented reality devices would require a high degree of accuracy, else holograms would be notably less immersive. Other tasks may only require an accuracy that can differentiate between rooms and floors; for example, when data linking information about the current room. Reliability is a measure of how often the localization is correct, and how fast the location tracking can drift off again. As wrong pose information can cause crucial errors (for example, a worker is shown data of the wrong object and acts on false information), reliability should be treated as the most important measure for critical tasks. Performance measures the computational efficiency and scalability of used algorithms. Computational efficiency is simply the amount of time an algorithm needs for localization. Scalability describes the complexity of the algorithm, and how execution time is impacted when in larger environments or when more data is given. Oftentimes, accuracy/reliability and performance are a tradeoff. An algorithm may provide more accurate or more reliable poses if given more time or computational power. Low performance may break immersion, limit user interaction, or drain a device's battery. Applications intended for users should pay attention to the performance of the localization. The cost not only includes the cost of performing a localization but also the setup and running cost of the necessary equipment integrated in the building. Localization approaches with markers or beacons especially require a considerable deployment effort. According to these factors, we will present a non-exhaustive list of localization approaches that are currently used or are in development.

Visual markers, such as QR codes or AR markers, are the most straightforward approach for localization. They provide a good accuracy and reliability to the localization, while requiring relatively cheap hardware and low computational power. Even with a large number of markers in a building, the performance scales well. The disadvantage of this type of localization is that a marker always needs to stay in view of the device. Either a large number of markers have to be distributed around the building, or localization can only be achieved at certain points (Hübner *et al.*, 2018). Additionally, there is a deployment effort when implementing this localization method, as markers need to be positioned around the building. The most common localization methods currently available rely on initial localization through markers, with pose updates through dead reckoning when the marker is not visible. This means that the accuracy and the reliability decrease over time after the first localization and have to be relocalized frequently.

Localization through beacons avoids the problem of always requiring visual input by using radio signals for positioning. Beacon localization through RFID or Wi-Fi works by measuring signal strength and using trilateration to find the position of the device. To achieve this, multiple beacons have to be spread around the building. For a 3D position, the device always needs to be in reach of at least three beacons. The range of beacons depends on the technology used. Passive RFID on ultra high frequency can be read from distances of 6 m, while active RFID and Wi-Fi beacons can reach up to 50 m. Of course, occlusions and reflections caused by walls and floors are detrimental to accuracy and reliability (Zhou and Shi, 2009). Thus, beacons should only be used when a rough location suffices; for example, for classification tasks (Ashfahani *et al.*, 2019).

Another method of localization is point-cloud matching. A point cloud is a collection of three-dimensional points that can be generated by using depth cameras, depth estimation algorithms, or laser scanners. Point clouds usually have a high accuracy and store a large amount of data. While point-cloud alignment allows high accuracy localization in the range of millimeters, the process itself has to compute on a large dataset, making computation extremely slow. Point-cloud localization is therefore only viable for non-real time applications (Bueno *et al.*, 2018).

Floor plan matching is similar to point-cloud matching as in that the surrounding environment is captured and mapped against the Digital Twin model. To combat the computational load of point-cloud algorithms, the data from depth sensors is simplified to two-dimensional floor plans which are matched with a floor plan of the Digital Twin (Herbers and König, 2019). This improves the performance significantly while still maintaining a high accuracy, but floor-plan matching does still have performance problems when covering large areas. Furthermore, floor-plan algorithms are more susceptible to self-similarities in buildings. If the layout of two floors is exactly the same, a floor-plan matching algorithm cannot differentiate between them. Reliability can thus only be assured in buildings without self-similarities.

A current focus of research is in machine learning algorithms for localization. Pose regression from visual input is able to recognize learned places from a trained model (Lowry, 2016). In most approaches in current research, a convolutional neural network is trained on images of the inside of a building with given poses. The training process is currently non-trivial and requires capturing the already localized images to train the model. Accuracy of current approaches is around a few meters, and also suffers when encountering self-similarities in a building (just as a human can mistake a room for another) (Walch *et al.*, 2017). Combining this approach with Digital Twins would allow for training the neural networks on a virtual model (Acharya *et al.*, 2019). The advantage of localization through machine learning is that executing a trained neural net is extremely performant, even in large environments. Modern smartphone hardware is being specialized for utilizing machine learning for complicated tasks, making it the most promising localization technology for the future.

Which type of localization is appropriate depends on the above-stated requirements: accuracy, reliability, available hardware performance, and cost of installation and running. Table 6.2 lists the comparisons between localization methods. Of course, multiple approaches can be combined through sensor fusion to mitigate some of the downsides of each approach. Combining sensor approaches with machine learning may be able to provide cost efficient localization with good accuracy and reliability in the future.

6.7 Case Studies

In this section three usable cases are presented which were developed within the research on AR technology at the Ruhr-Universität Bochum. The first case comprises an AR-supported teaching environment in which a teacher can link instructions and explanations with a laboratory device to teach the students how to use the device. Such a teaching environment could also be used to explain, to maintenance workers, how to maintain parts of a cognitive building using smart maintenance. The second case study explains how a maintenance process can be digitized step by step. The case study presented here deals with an AR-assistance system that obtains information for maintenance from a P&I (piping and instrumentation) diagram. This is especially interesting for the area of cognitive buildings, since the development of digitized maintenance processes is essential for such a structure. In the last case study, the Augmented Reality Content Management System, or ARCMS for short, is presented. This is a system in which content is created and managed for the use of AR systems, which is later deployed in various applications.

6.7.1 Collaborative Augmented Reality

The vision behind the collaborative framework for augmented reality scenarios in engineering education is shown in Fig. 6.8. Given that both the teacher and the student handle smart devices, which are connected to the same network, the teacher makes use of her knowledge and preparation to give the student insight into the laboratory machine's working. With the ability to control the focus of the student's augmented reality overlay, the teacher creates step-by-step instructions and explanations which are based on sensory data that the students can explore themselves. The approach is not aimed at replacing all interaction in the laboratory with app-based communication, but to augment the argumentation of teachers with interactive elements. The students are therefore enabled to find individual approaches towards the learning scenario, take different perspectives of the apparatus into account, and try several virtual steps before the experiment is undertaken by them. They follow a real experiment, which

Table 6.2: Comparison of Localization Methods

Localization method	Accuracy	Reliability	Performance	Cost
Visual markers (QR-codes, etc.)	High accuracy when in view	Only reliable when directly in view	Good	Medium initial cost, no running cost
Beacons	Low to medium accuracy	Reliable when in range	Good	High initial cost, low or no running costs
Point-cloud matching	High	Good	Low	Low
Floor plan matching	High	Only reliable if the floor plan is distinct enough	Low scalability	Almost none
Machine learning	Low (in current research)	Depending on self-similarity in a building	Good	Cost of training a model, no running costs

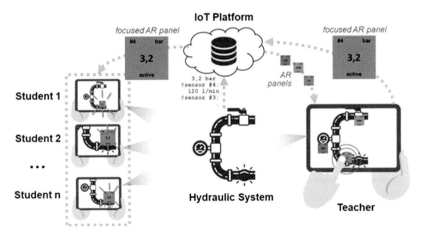

Fig. 6.8: Framework for augmented reality scenarios in engineering education
(Neges *et al.*, 2019)

is performed by a teacher, to identify/suggest the next steps while the teacher can steer the ideas and instantly show the experiments' behavior. This way, learning gets more individualised by using digital technologies. The approach synchronizes the views of both the teacher and her students while displaying sensor data. An Internet of Things (IoT) platform provides ubiquitous connectivity between smart devices, which makes them ideal for hosting data for distributed applications as in this case.

The usage of augmented reality in higher education is still relatively rare, but several approaches exist that make use of either motivational aspects of AR or support the cooperative aspect of diagnosis with digital information. The concept consists of the following components:

- the apparatus (a hydraulic system) as the center of the experiment;
- the instructor who prepares her lessons by creating and positioning AR panels;
- multiple learning students;
- the smart device applications for teachers and students as well as
- an IoT platform that manages the sensor data and synchronizes all connected clients.

The connections and dependencies between these components are shown in Fig. 6.8. In an initial phase, the apparatus needs to be connected to the IoT platform, so that the available (and relevant) sensory data is present, up to date and defined by type before the teacher plans the experiment. When planning, the teacher creates type-specific AR panels (orange panels represent pressure, blue panels represent flow rate, etc.) and places them in the learning environment via touch gestures. In the lecture itself, the teacher activates her currently relevant panel by gesture.

There are two options given at this moment – the first is to reveal the hidden panel to the students in addition to the ones that have been unlocked before; the second option is to show only the last activated panel. The goal of a graph-based approach is to create a dynamic model as a single-source-of-truth for maintenance processes. It allows manipulation from either the planning side or the operational side by means of integration into the regular workflows of the involved stakeholders, such as engineers and technicians.

6.7.2 Round-trip Engineering Between P&I Diagrams and Augmented Reality Work Instructions

The goal of a graph-based approach is to create a dynamic model as a single-source-of-truth for maintenance processes. It allows manipulation from either the planning side or the operational side by means of integration into the regular workflows of the involved stakeholders, such as engineers and technicians.

Models, as a depiction of reality, must be convincing enough in regard to the situation and problem position. When looking at hydraulic systems and possible ways to depict them in a formalized model, it becomes clear that one has to look for certain things and their interconnections. In both cases, not only is the thing or connection itself relevant, but also its context and type. It is thereby possible to build classes and hierarchies. The first assertion is that hydraulic systems in general, and their representation in P&I diagrams in particular, can be interpreted as a graph, whereby the afore-mentioned things are represented through nodes and their logical or physical connections are represented by edges.

The approach for the round-trip engineering enabled graph-based model for the provision of context-sensitive augmented reality work instructions consists of several components and steps. As the maintenance technician is not only the recipient of information for the assisted maintenance process, but also the author of data through her manual labor, the concept allows feedback data to flow back into the graph-based model and from there to the designated engineering documents. To illustrate the approach, Fig. 6.9 shows the idea and its interconnected components, actions, and stakeholders.

The entry point to this concept is step 1.a, the creation of a P&I diagram by an engineer. At this stage, the incorporation of the diagram into the feedback cycle concept does not affect the engineer. Step 1.b is optional, as the maintenance manager may authorize an Action Request or Engineering Change Request (ECR) to physically change the machinery. Such a change may include the addition of a sensor or component or the replacement of such.

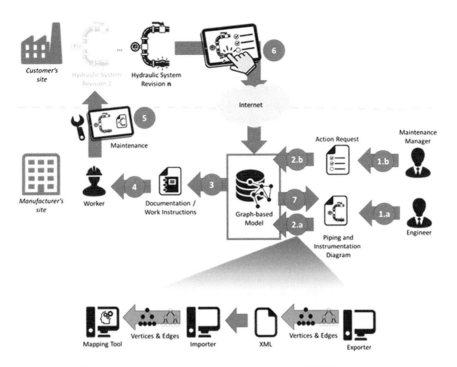

Fig. 6.9: Concept of the round-trip engineering for AR (Neges *et al.*, 2017)

The second step in Fig. 6.9 consists of the creation of an intermediate file with the necessary data to be imported into the graph-based model from the P&I diagram. To achieve this, step two consists of several phases to realize the interfacing between graphical representation and usable data in a graph database. For this, export from the graphical representation, the proprietary authoring software needs either interfaces, add-on-capabilities, or an inbuilt programming module. Either of these possibilities enables the parsing of the data behind the graphical representation for the extraction of edge and vertex classes and instances of edges and vertices. In the context of P&I diagrams, this means the registration of edge classes for pipes or control wire; for example, along with the corresponding registration of individual pipes with their respective start and end points as well as further meta data. The same principle applies to vertex classes and vertices. All components in the P&I diagram are analyzed to extract the used classes of components, such as valves, flow meters, or tanks. Subsequently, the exporter catalogs the individual components with their specific data.

On the side of the graph-based model, the corresponding importer for the XML scheme loads the data and compares it to the data already present. Using the mapping tool, the engineer either can map classes for vertices and edges to existing classes in the graph-based model or mark them as new to create a new cluster of elements.

The third step in Fig. 6.9 includes usage of the author's algorithms to conduct an analysis of the machine at hand with the graph-based model. The fourth step consists of the analysis through the graph-based model that can trigger warnings and give guidance to reestablish a desired operation state. The fifth step is the physical maintenance process in itself with added support through the aforementioned assistance system using augmented reality visualization. This proven approach offers service technicians an interactive, intuitive, and documentable way to maintain machinery. To realize context awareness, a smart device application uses graph-based algorithms to guide the technician to achieve the machine's desired operating condition after evaluating the current state of the machine.

The sixth step is the checkout process in which the technician confirms her actions, again by using the smart device application. By triggering the prepared actions through the graph-based model, the assistant application uploads the feedback data. As the mapping between model and engineering documents is bidirectional, the seventh step then uses the freshly uploaded data to detect possible inconsistencies between the engineering document and the graph-based model. The feedback assistant then uses the delta between the outdated representation in the engineering document and the more current data in the model to visualize which changes the engineer has to make (Fig. 6.10).

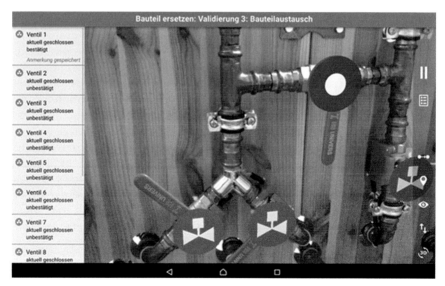

Fig. 6.10: Dynamic maintenance app in use (Wolf, 2019)

6.7.3 The Augmented Reality Content Management System

One of the major roadblocks for a widespread usage of augmented reality is content creation. To help the spread of augmented reality (AR) in an education environment, an Augmented Reality Content Management System (ARCMS) was created by the chair for digital engineering at Ruhr-Universität Bochum. AR has the potential to make paper-based manuals superfluous as the technical hurdles get lower and lower. Tablet computers offer a better performance for less money and the necessary software kits are built directly by the manufacturers. But content creation remains to be a problem reserved for experts. Most systems require high-level programming languages. Geometries must be created in special tools, CAD data has to be post-processed to create impressive AR scenarios. But the real benefit of AR-based instructions is the presentation of context and location-sensitive information.

Creating AR instructions is a two-step process – first, the task gets split into small, actionable subtasks that are entered in a web interface. For each subtask, texts, images, links and safety symbols are added to give the user enough information to successfully finish the task. AR only plays a role in the second step. For each step, an AR hint gets placed at the real machinery using a tablet application. The user can choose from a small library of prepared hints, including safety symbols and a variety of arrows. The selected position gets saved with the rest of the content in a central database. The same tablet application is used by the end-user to view the whole instruction. There, the placed AR hints are shown on the tablet together with the description, links, and images. The instruction then helps the user to solve the task step by step.

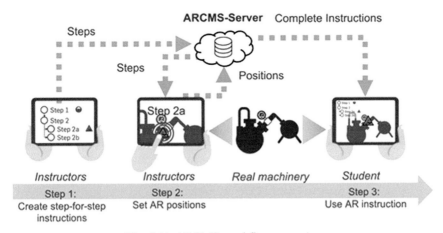

Fig. 6.11: ARCMS workflow overview

The system is used in education at the Ruhr-Universität Bochum. At chairs from the Department of Process Engineering as well as chairs from the Department of Material Science, AR-based instructions are used to guide students through laboratory experiments. Conducted studies prove the usefulness and ease-of-use of the system to both educators as well as learners (Fig. 6.12).

Fig. 6.12: ARCMS app in use

The ARCMS should be further propagated by offering training to educators and implementing a direct connection to the E-Learning system being used. Because of the modularized concept, the ARCMS can be used in completely unrelated fields; for example, at the Center for Medicine Didactics, the system is being used by medical students in the 'Skills Lab'. Together with the Chair for History Didactics, the system explains the historic exhibits.

6.8 Conclusion

In this chapter we have shown how Digital Twins can be employed to help keep track of material and machines used in the construction of a real building. Additionally, approaches to linking a Digital Twin to its physical counterpart are presented. We have also shown how augmented reality can be used in maintenance. In combining these technologies, a smart maintenance system can be achieved which helps on-site maintenance personnel in multiple ways. Data linking can be used to improve predictive maintenance in cognitive buildings, utilizing all sensory data in a holistic way. Functional linking allows a maintenance worker to interact with the building through AR devices. Pose linking the Digital Twin to an AR device allows for live overlays of data, which can facilitate multiple improvements in the maintenance process.

Furthermore, it is important that after maintenance is performed or anything in the building has changed, the Digital Twin is updated. When kept up to date, either through manual input or the smart maintenance system itself, the Digital Twin can be the single source of truth for all information about the real building it represents. With this knowledge, a cognitive building could tap into its smart maintenance system to monitor building components through Digital Twin. The smart maintenance system could automatically schedule maintenance tasks, either for preventive maintenance or reactive maintenance. In some cases, the intelligent maintenance system could even completely automate the maintenance process, for example, by using robotics. These concepts are still in the early stages of development and are far from being practical applications. Nevertheless, smart maintenance is the oil that keeps a cognitive building running. In other words, intelligent maintenance is an integral part of designing and operating a cognitive building.

References

Abramovici, M., Krebs, A. and Wolf, M. (2014). Approach to ubiquitous support for maintenance and repair jobs utilizing smart devices. *In*: Tools and Methods of Competitive Engineering: Digital Proceedings of the Tenth International Symposium on Tools and Methods of Competitive Engineering – TMCE, 2014, S. 19-23.

Acharya, D., Khoshelham, K. and Winter, S. (2019). BIM-PoseNet: Indoor camera localisation using a 3D indoor model and deep learning from synthetic images. *ISPRS J. Photogramm. Remote Sens.*, Bd. 150: 245-258, März 2019. doi: 10.1016/j.isprsjprs.2019.02.020

Ashfahani, A., Pratama, M., Lughofer, E., Cai, Q. and Sheng, H. (2019). An Online RFID Localization in the Manufacturing Shopfloor. *In*: Predictive Maintenance in Dynamic Systems: Advanced Methods, Decision Support Tools and Real-World Applications, E. Lughofer und M. Sayed-Mouchaweh, Hrsg. Cham: Springer International Publishing, 2019, S. 287-309.

Azuma, R.T. (1997). A survey of augmented reality. *Presence Teleoperators Virtual Environ.*, Bd. 6, Nr. 4, S. 355-385, Aug. 1997. doi: 10.1162/pres.1997.6.4.355

Azuma, R.T. (2016). The most important challenge facing augmented reality. *Presence*, Bd. 25, Nr. 3, S. 234-238. doi: 10.1162/PRES_a_00264

Baumeister, J., Ssin, S.Y., ElSayed, N.A., Dorrian, J., Webb, D.P., Walsh, J.A. and Thomas, B.H. (2017). Cognitive cost of using augmented reality displays. *IEEE Trans. Vis. Comput. Graph.*, Bd. 23, Nr. 11, S. 2378-2388, Nov. 2017. doi: 10.1109/TVCG.2017.2735098

Billinghurst, M., Clark, A. and Lee, G. (2015). A Survey of augmented reality. *Found. Trends Human–Computer Interact.*, Bd. 8, Nr. 2–3, S. 73-272. doi: 10.1561/1100000049

Bokrantz, J., Skoogh, A., Berlin, C., Wuest, T. and Stahre, J. (2020). Smart maintenance: An empirically grounded conceptualization. Int. *J. Prod. Econ.*, Bd. 223, S. 107534. doi: 10.1016/j.ijpe.2019.107534

Borrmann, A., König, M., Koch, C. and Beetz, J. (2018). Building Information Modeling: What? *In*: Building Information Modeling, pp. 4-11, Springer, Cham.

Borrmann, A., König, M., Koch, C. and Beetz, J. (2018). Building Information Modeling: Why? *In*: Building Information Modeling, pp. 2-3, Springer, Cham.

Borrmann, A., König, C. Koch, C. and Beetz, J. Hrsg (2018). Industry foundation classes: a standardized data model for the vendor-neutral exchange of digital building models. *In*: Building Information Modeling: Technology Foundations and Industry Practice, Springer International Publishing, S. 81-86.

Boschert, S., Heinrich, C. and Rosen, R. (2018). Next generation Digital Twin. *In*: Proc. TMCE, pp. 209-218, Las Palmas de Gran Canaria, Spain.

Bueno, M., Bosché, F., González-Jorge, H., Martínez-Sánchez, J. and Arias, P. (2018). 4-Plane congruent sets for automatic registration of as-is 3D point clouds with 3D BIM models. *Autom. Constr.*, Bd. 89, S. 120-134. doi: 10.1016/j.autcon.2018.01.014

Chen, W., Cheng, J.C. and Tan, Y. (2019). BIM- and IoT-based Data-Driven decision support system for predictive maintenance of building facilities. *In*: Innovative Production and Construction, World Scientific, S. 429-447.

Deutsche Akademie der Technikwissenschaften (Acatech) (2015). Smart Maintenance für Smart Factories –Mit intelligenter Instandhaltung die Industrie 4.0 vorantreiben, München: Utz, Herbert (acatech POSITION).

DIN 31051 (2019). *Fundamentals of Maintenance*, June 2019.

DIN 13306 (2018). *Maintenance – Maintenance Terminology*, Feb. 2018.

DIN EN 15341 (2007). *Maintenance – Maintenance Key Performance Indicators*, 2007.

DIN EN 15628 (2014). *Maintenance – Qualification of Maintenance Personnel*, Okt. 2014.

Dini, G. and Dalle Mura, M. (2015). Application of augmented reality techniques in through-life engineering services. *Procedia CIRP*, Bd. 38, S. 14-23. doi: 10.1016/J.PROCIR.2015.07.044

Dörner, R., Broll, W., Grimm, P. and Jung, B. Hrsg (2013). Virtual und Augmented Reality (VR/AR): *Grundlagen und Methoden der Virtuellen und Augmentierten Realität.*, Springer Vieweg.

Fuller, A., Fan, Z., Day, C. and Barlow, C. (2020). Digital Twin: Enabling Technologies, Challenges and Open Research. ArXiv191101276 Cs, Mai 2020. doi: 10.1109/ACCESS.2020.2998358

Gnoni, M.G. and Saleh, J.H. (2017). Near-miss management systems and observability-in-depth: Handling safety incidents and accident precursors in light of safety principles. *Saf. Sci.*, Bd. 91, S. 154-167, 2017. doi: 10.1016/j.ssci.2016.08.012

Grosse, C.U. (2019). Monitoring and Inspection Techniques Supporting a Digital Twin Concept in Civil Engineering. *In*: Proceedings of the Fifth International

Conference on Sustainable Construction Materials and Technologies, Kingston University, London, Juli 2019, S. 16, [Online], Verfügbar unter: http://www.claisse.info/Proceedings.htm

Herbers, P. and König, M. (2019). Indoor localization for augmented reality devices using BIM, point clouds, and template matching. *Appl. Sci.*, Bd. 9, Nr. 20, S. 4260. doi: 10.3390/app9204260

Hou, L., Wang, X. and Truijens, M. (2015). Using augmented reality to facilitate piping assembly: An experiment-based evaluation. *J. Comput. Civ. Eng.*, Bd. 29, Nr. 1, S. 05014007. doi: 10.1061/(ASCE)CP.1943-5487.0000344

Hübner, P., Weinmann, M. and Wursthorn, S. (2018). Marker-based localization of the microsoft hololens in building models. *ISPRS – Int. Arch. Photogramm. Remote Sens. Spat. Inf. Sci.*, Bd. XLII-1, S. 195-202, Sep. 2018. doi: 10.5194/isprs-archives-XLII-1-195-2018

Koch, C., Neges, M., König, M. and Abramovici, M. (2014). Natural markers for augmented reality-based indoor navigation and facility maintenance. *Autom. Constr.*, Bd. 48, S. 18-30. doi: 10.1016/J.AUTCON.2014.08.009

Kuhn, A. and Bandow, G. (2009). Industrielles Servicemanagement – Garant für unternehmerischen Erfolg. Fraunhofer-Institut für Materialfluss und Logistik.

Kuhn, A., Schuh, G. and Stahl, B. (2006) Nachhaltige Instandhaltung. Trends, Potenziale und Handlungsfelder nachhaltiger Instandhaltung. VDMA-Verlag, Frankfurt a.M., 2006.

Lamberti, F., Manuri, F., Sanna, A., Paravati, G., Pezzolla, P. and Montuschi, P. (2014). Challenges, opportunities and future trends of emerging techniques for augmented reality-based maintenance. *IEEE Trans. Emerg. Top. Comput.*, Bd. 2, Nr. 4, S. 411-421, Dez. 2014. doi: 10.1109/TETC.2014.2368833

Lamkin, P. (2017). Virtual Reality Headset Sales Hit 1 Million. *Forbes*, Nov. 30, 2017. https://www.forbes.com/sites/paullamkin/2017/11/30/virtual-reality-headset-sales-hit-1-million/ (zugegriffen Apr. 29, 2020)

Lee, H.H.Y. and Scott, D. (2009). Overview of maintenance strategy, acceptable maintenance standard and resources from a building maintenance operation perspective. *J. Build. Apprais.*, Bd. 4, Nr. 4, S. 269-278. doi: 10.1057/jba.2008.46

Lowry, S., Sünderhauf, N., Newman, P., Leonard, J.J., Cox, D., Corke, P. and Milford, M.J. (2015). Visual place recognition: A survey. *IEEE Trans. Robot.*, Bd. 32, Nr. 1, S. 1-19, Feb. 2016. doi: 10.1109/TRO.2015.2496823

Lughofer, E. and Sayed-Mouchaweh, M. Hrsg (2019). *Predictive Maintenance in Dynamic Systems: Advanced Methods, Decision Support Tools and Real-World Applications*. Cham: Springer International Publishing.

May, M. Hrsg (2018). Digital Twins. *In*: CAFM-Handbuch: Digitalisierung im Facility Management erfolgreich einsetzen, Wiesbaden: Springer Fachmedien Wiesbaden, S. 346, 347.

Milgram, P., Takemura, H., Utsumi, A. and Kishino, F. (1994). Augmented reality: A class of displays on the reality-virtuality continuum. *In*: Telemanipulator and Telepresence Technologies, 1994, Bd. 2351. doi: 10.1117/12.197321

Mobley, R.K. (2002). *An Introduction to Predictive Maintenance*. Elsevier.

Neges, M., Wolf, M., Kuska, R. and Frerich, S. (2019). Framework for augmented reality scenarios in engineering education'. *In*: Smart Industry & Smart Education, Cham, S. 620-626. doi: 10.1007/978-3-319-95678-7_68

Neges, M., Wolf, M. and Abramovici, M. (2017). Enabling round-trip engineering between &I diagrams and augmented reality work instructions in maintenance processes utilizing graph-based modelling. *In*: Intelligent Systems in Production Engineering and Maintenance – ISPEM 2017, Cham, 2018, S. 33-42. doi: 10.1007/978-3-319-64465-3_4

Pašek, J. and Sojková, V. (2018). Facility management of smart buildings. *Int. Rev. Appl. Sci. Eng. IRASE*, Bd. 9, Nr. 2, [online]. Verfügbar unter. https:// akjournals.com/view/journals/1848/9/2/article-p181.xml

Pawellek, G. *Integrierte Instandhaltung und Ersatzteillogistik*, Berlin, Heidelberg: Springer, 2013.

Rekimoto, J. and Nagao, K. (1995). The world through the computer: Computer augmented interaction with real world environments. *In*: Proceedings of the 8th Annual ACM Symposium on User Interface and Software Technology, Pittsburgh, Pennsylvania, USA, Dez. 1995, S. 29-36. doi: 10.1145/215585.215639

Sanpechuda, T. and Kovavisaruch, L. (2008). A review of RFID localization: Applications and techniques. *In*: 5th International Conference on Electrical Engineering/Electronics, Computer, Telecommunications and Information Technology, Mai 2008, Bd. 2, S. 769-772. doi: 10.1109/ECTICON.2008.4600544

Schlick, C.M., Moser, K. and Schenk, M. Hrsg (Eds.) (2014). Unterstützung von Lernprozessen bei Montageaufgaben. *In*: Flexible Produktionskapazität innovativ managen: Handlungsempfehlungen für die flexible Gestaltung von Produktionssystemen in kleinen und mittleren Unternehmen, Springer, Vieweg.

Speicher, M., Hall, B.D. and Nebeling, M. (2019). What is Mixed Reality? *In*: Proceedings of the 2019 CHI Conference on Human Factors in Computing Systems, Glasgow, Scotland UK, Mai 2019, S. 1-15. doi: 10.1145/3290605.3300767

Strunz, M. (2012). Gegenstand, Ziele und Entwicklung betrieblicher Instandhaltung. *In*: Instandhaltung: Grundlagen - Strategien - Werkstätten, Berlin, Heidelberg: Springer Berlin Heidelberg, 2012, S. 1-35.

SuperData (2019). In Review: Digital Games and Interactive Media [online]. Verfügbar unter. https://www.superdataresearch.com/reports/2019-year-in-review

Susto, G.A., Schirru, A., Pampuri, S., McLoone, S. and Beghi, A. (2014). Machine learning for predictive maintenance: A multiple classifier approach. *IEEE Trans. Ind. Inform.*, Bd. 11, Nr. 3, S. 812-820. doi: 10.1109/TII.2014.2349359

Tönnis, M. (2010). *Augmented Reality: Einblicke in die Erweiterte Realität*. Berlin Heidelberg: Springer-Verlag.

Van Krevelen, D.W.F. and Poelman, R. (2010). A survey of augmented reality technologies, applications and limitations. *Int. J. Virtual Real.*, Bd. 9, Nr. 2, S. 1-20. doi: 10.20870/IJVR.2010.9.2.2767

Walch, F., Hazirbas, C., Leal-Taixe, L., Sattler, T., Hilsenbeck, S. and Cremers, D. (2017). Image-based localization using LSTMs for structured feature correlation. *In*: 2017 IEEE International Conference on Computer Vision (ICCV), Venice, Okt. 2017, S. 627-637. doi: 10.1109/ICCV.2017.75

Werner, A., Zimmermann, N. and Lentes, J. (2019). Approach for a holistic predictive maintenance strategy by incorporating a Digital Twin. *Procedia Manuf.*, Bd. 39, S. 1743-1751. doi: 10.1016/j.promfg.2020.01.265

Wiedemann, M. and Wolff, D. (2013). Instandhaltung – Handlungsfelder zur Optimierung der softwaretechnischen Unterstützung im Kontext von Industrie 4.0. *ZWF*, Bd. 108, Nr. 11, S. 805-808, 2013; doi: 10.3139/104.013118

Wolf, M. (2019). Smartes Instandhaltungs-Assistenzsystem für verfahrenstechnische Anlagen in Industrie 4.0. *Shaker*.

Zapata, P., Cárdenas, C. and Lozano, N. (2019). Building information modeling 5D and earned value management methodologies integration through a computational tool. *Revista Ingeniería de Construcción*, 33(3): 263-278.

Zhou, J. and Shi, J. (2009). RFID localization algorithms and applications – A review. *J. Intell. Manuf.*, Bd. 20, Nr. 6, S. 695-707. doi: 10.1007/s10845-008-0158-5

Blockchain: Technologies for Facilitating Cyber-Physical Security in Smart Built Environment

Jong Han Yoon[1], Xinghua Gao[2] and Pardis Pishdad-Bozorgi[1*]

[1] School of Building Construction, Georgia Institute of Technology, USA
[2] Myers-Lawson School of Construction, Virginia Polytechnic Institute and State University, USA

7.1 Introduction

In smart built environments, people's lives digitally connect with city services, buildings, infrastructures, and utilities through information and communication technologies (ICTs) (Gračanin *et al.*, 2015). Typically, a smart built environment may consist of components, such as smart cities, smart grids, smart buildings, and smart devices.

In a smart city, the digital connection between people's lives and city services will provide people with a better quality of life, a better social life, and energy sustainability through optimized city services management based on the analysis of the information on digital connections (Zubizarreta *et al.*, 2016). These connections are enabled by ICTs, such as the Internet of Things (IoT) technology and cloud technology (Baig *et al.*, 2017; Zang *et al.*, 2017). The technologies also make electric power grids more energy-efficient and reliable by collecting and providing energy-associated information, thus informing generators' and consumers' decisions for an optimal energy management (Gunduz and Das, 2020). These new technologies-enabled grids are typically called 'smart grids'. Compared with traditional power grids, the smart grids provide more reliable, secure, and economical transmission of electricity by protecting,

---
*Corresponding author: pardis.pishdad@gatech.edu

monitoring, analyzing, and controlling the electricity transmission processes (Gunduz and Das, 2020). The technologies (i.e. IoT and cloud) also empower smart buildings, which enable building occupants to have more convenient and comfortable lives because the building systems are optimally operated and controlled with advanced technologies, such as Big Data engineering and IoT (Jia *et al.*, 2019); for instance, Building Automation System (BAS), which is a complex network-based distributed control system enabling communication and cooperation of electrical/ mechanical subsystems, provides optimized control of HVAC, lighting, and air humidity and also manages building security and safety (Wang *et al.*, 2015). Smart devices assist other components of the smart built environments (i.e. smart city, smart grid, and smart building) by enabling them to collect and process associated data from a physical world and connecting the smart built environments so that they can exchange data for their operations (Silverio-Fernández *et al.*, 2018).

Since the ICT applications in the smart built environments are continually increasing and evolving, the potential benefits that the smart built environments can provide are also growing. However, coupled with the advantages of smart built environments are some concerning disadvantages, like cyber attacks. With an increase usage of ICT applications, the smart built environments are increasingly subjected to cyber attacks. The threat of cyber attack to smart built environments has been investigated by many researchers (Overman *et al.*, 2011; Wang *et al.*, 2015; Paridari *et al.*, 2016; Baig *et al.*, 2017; Minoli *et al.*, 2017). In smart built environments, cyber attacks will be more serious because they directly impact people's lives. In this setting, the cyber security risks and physical security risks are resulting into Cyber-Physical Security risks. Such risks can be substantial when huge quantities of data, collected and processed by ICT to operate smart built environments, are attacked (AlDairi, 2017). In the case of smart buildings, the private data of occupants, such as their location or behavior patterns, can be stolen from ICT platforms and exploited with criminal intent (Do *et al.*, 2018); for example, criminals may disable a smart building's security system via cyber attack, taking over the assess control, which endangers the occupants. In the context of Smart Grid, cyber attack can lead to breaks in national security, disruption of public order, and loss of life or large-scale economic damage (Vitunskaite *et al.*, 2019; Gunduz and Das, 2020).

Blockchain technology has the potential to mitigate the threats and risks of cyber attacks against smart built environments. The blockchain can generate a data platform for IoT systems in which data theft and tampering are basically impossible with its hash function, consensus mechanism, and capability of distributing the data transaction records to every participant (Samaniego and Deters, 2016). Even though these benefits can enhance cyber security of smart built environments, a limited number of studies demonstrate that the blockchain-based data sharing platform can enhance the security for the smart built environments (Biswas and Muthukkumarasamy, 2016; Aung and Tantidham, 2017; Dorri *et al.*, 2017; Mylrea and Gourisetti, 2017; Pop *et al.*, 2018; Qu *et al.*, 2018; Minoli, 2019; Agung and Handayani, 2020; Makhdoom *et al.*, 2020, Qashlan *et al.*, 2020) {Gunduz, 2020 #152}.

In this chapter, through literature reviews, the authors investigate the potential of blockchain in mitigating cyber-physical risks in the smart built environments. The authors highlight key aspects of the Cyber-Physical Security and risks, discuss potential of blockchain for addressing them, and propose future research directions to enlighten new researchers.

7.2 Cyber-physical Risks in Smart Built Environments

This section discusses the definition of each layer of smart built environments and reviews the literature regarding the cyber-physical risks according to the type. The smart built environments are characterized with separate layers: Smart city, smart infrastructure, smart buildings, and smart devices. Even though the smart city consists of numerous smart components (e.g. smart infrastructure, smart transportation, smart governance, smart services, etc.), in this section, the authors address the built-environment components, such as smart infrastructures, buildings, and devices. In addition, the authors focus on the energy network of smart infrastructures, which is the smart grid. These components are interconnected with each other through ICTs, such as IoT devices, cloud database, and Big Data. This interconnection enables the operation of the smart city (Mohanty *et al.*, 2016) (Fig. 7.1). Because each layer has its specific cyber-physical risks, focusing on the unique characteristics of each layer will clarify the differences among the reviewed works.

7.2.1 What is Smart City?

Although a great number of studies define what a smart city is, the authors refer to the definition that highlights the connection between smart systems and people's lives through Cyber-Physical Systems (CPSs). CPSs use internet network and integrating computational technologies to access data, process data, and impact surrounding physical environment (Monostori, 2018).

Smart cities are resilient, facilitate mobility, add efficiencies, conserve energy, improve the quality of air and water, identify and solve problems quickly, recover rapidly from disasters, collect data to make better decisions, deploy resources effectively, and share data to enable

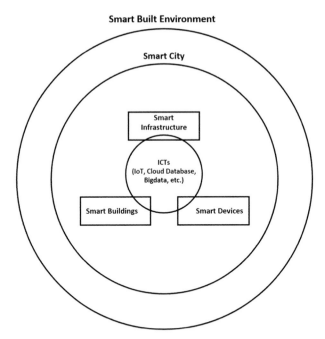

Fig. 7.1: Layers of smart built environment (adapted from Mohanty *et al.*, 2016)

collaboration across entities and domains by infusing information into the cities physical infrastructure (Nam and Pardo, 2011). By utilizing the collected data and information-infused infrastructure, the cities can promote the quality of people's lives in the city in a more timely, effective, and energy-efficient way (Chamoso *et al.*, 2020). Harrison *et al.* (2010) notes that in smart cities, the traditional concept of a physical city is extended to a virtual city, which utilizes ICT to physically impact people's lives. Not only are the ICT applications important for the definition of the smart city, but their impacts on the people's educational capital, social and relational capital, and environmental issues are also significant (Lombardi *et al.*, 2012). Marsal-Llacuna *et al.* (2015) state, "*The smart cities initiative seeks to improve urban performance by using data, information and Information Technologies (IT) to provide more efficient services to citizens, to monitor and optimize existing infrastructure, to increase collaboration between different economic actors and to encourage innovative business models in both the private and public sectors.*" Accordingly, in the smart city, people's quality of life, their decisions about their behavior, and their social life are directly linked to the ICT-applied city infrastructure, which means that cyber attacks on information in the infrastructure can result in severe physical attacks on the real lives of the people in the smart city.

7.2.2 Existing Cyber-Physical Security Risks and Solutions for Smart Cities

As the operation of smart cities heavily depends on ICTs, the cities are vulnerable to cyber attacks, which may lead to significant losses (Vitunskaite *et al.*, 2019). The cyber attacks against data that is processed and shared within the city, and any attack can cause problems closely related to people's lives, such as financial loss and private data theft (Chauhan *et al.*, 2016). Accordingly, many studies have investigated the cyber security of smart cities in order to prevent or mitigate cyber attacks against the cities.

Wang *et al.* (2015) have proposed a security threat assessment model that can be used to create risk mitigation strategies against cyber intrusions into smart city systems. To develop the assessment model, the authors established a threat model by collecting and analyzing security-associated information in the network and system architecture, operating systems and updates, components and configurations of applications, data and data storage, database schemas, services and roles, and encryptions and external dependencies. From the threat model, the authors generated a list of threats and associated risks, which were combined with an algorithm to calculate the threat factors. They then used these factors to assess the security threats and provide appropriate mitigation strategies. Wang *et al.* (2015) also verified that the mitigation strategies through threat assessment can improve the security of the smart city systems in case-based experiments. However, their study was limited to risk assessment and countermeasures based on assessment. They did not provide a preventive way to protect the smart city systems from cyber attacks. Furthermore, they concluded that their lifecycle threat assessment was too long as in that the second round of assessment and mitigations took three months, which meant the systems would have been exposed to the risks over an unrealistically long time period.

Another study for cyber security risks and solutions was conducted by Baig, Szewczyk *et al.* (2017) to investigate the security threats of data acquired from and stored in the smart city components (i.e. smart grids, BAS, unmanned aerial vehicles (UAV), smart vehicles), which utilize IoT sensors and a cloud-data-storage platform. To mitigate the threats, they highlighted the importance of digital forensic investigation on data. The digital forensic restores and analyzes the data for tracing the data use and identifying the evidence of criminal proceedings (Tully *et al.*, 2020). The digital forensic can be especially useful for improving the security of cloud storage, since the storage is connected to internet, which is vulnerable to cyber attacks (Chung *et al.*, 2012). Baig *et al.* (2017) found that the forensic investigation on smart cities components that use cloud storage can help the investigator deeply comprehend the cyber attacks and prevent the resulting negative event. One question that Baig *et al.* (2017) did not address, however, was how integrity of the data in the forensic investigation process can be confirmed and maintained so that it is not tampered with or leaked. Even though the researchers pointed out that the data should be kept tamper-proof for effective forensic investigations, they did not propose how to do this. One limitation of digital forensics is that it is not a proactive means for preventing cyber attacks to smart city components. More research to address proactive and preventive means is required.

The cyber security risks and solutions are also discussed by Zang *et al.* (2017). The article revealed that data can be damaged or corrupted in cloud storage of smart cities unless it has security-enhanced remote data auditing (RDA) protocols. These protocols protect cloud users from deceitful cloud servers, which may hide data loss incidents or delete data intentionally for illicit purposes (Zang *et al.*, 2017). The study improved RDA protocols compared to the RDA protocols of Sookhak (2015), which was vulnerable to two *replace*-and-*replay attacks*. A *replace attack* occurs when the server replaces damaged files with undamaged ones, and a *replay attack* occurs when the server deploys previous proofs and associated information to forge a valid proof for a new challenge (Zang *et al.*, 2017). The study verified that the proposed RDA protocol can effectively prevent those attacks, something which the previous RDA protocol could not do. Even though the improved protocol can protect cloud users from deceitful cloud servers, the cyber-physical attacks in smart cities can't be completely eliminated because the cyber attacks exist not only in the cloud servers but also in devices, networks, and computational resources (Delgado-Gomes *et al.*, 2015).

Some studies addressed the cyber-physical security risks from the aspects of risk cognition and information security management (ISM). Chatterjee *et al.* (2018) quantitatively verified that it is the risk cognition on the cyber security of smart cities that motivates people to use preventive technology against cyber threats in cities. The study also found that the cognition can be effectively achieved through social media, word of mouth, and official organization (e.g. banks, post offices, and financial institutions). These findings emphasize that the awareness of cyber attacks is important for people to prevent them and improve security by providing effective ways to increase awareness. Hasbini *et al.* (2018) investigated the role of ISM in smart city organizations. The study identified organizational factors, which enhance ISM in smart cities, highlighting the importance of information security governance. The identified factors can be utilized to establish an effective organization for smart cities to enhance information security.

Laufs *et al.* (2020) pointed out that the security and privacy issues of smart city systems are relatively neglected when considering that smart technologies increasingly address urban challenges directly affecting people's lives. The authors asserted that the privacy rights and data protection should be considered in the smart city planning processes. Even though the smart technologies generate more efficient city services or effectively mitigate crimes, the people might feel less secure when there is a chance for an unpermitted entity to control their life by using the technologies or to steal and exploit their private data (Laufs *et al.*, 2020). While afore-mentioned several articles studied cyber security risks and countermeasures in smart cities, more future research on data ownership and privacy rights are still required.

7.2.3 What is Smart Grid?

The smart grid is a power network enabling a two-way delivery of energy and information, which allows the power industry to optimize energy delivery and help consumers to optimally manage their energy usage (Delgado-Gomes *et al.*, 2015; Dileep, 2020; Vaccaro *et al.*, 2020). According to the National Institute of Standards and Technology (NIST), the smart grid can be established when the electrical grid domains, such as the generators, distributors, consumers, and operators, are connected and communicate with one another to efficiently deliver sustainable, economic, and secure electricity (SmartGrids, 2012). In this setting, the exchange of data or information among the domains should be secured and transparent so that each domain utilizes them in the intended way to provide smart power services (Vaccaro *et al.*, 2020). The operation of a smart grid relies on the information exchange between the energy provider and consumer through the energy data collection and process (Raut *et al.*, 2016). Deploying the data, the smart grid maximizes energy efficiency and minimizes energy loss (Mo *et al.*, 2011; Kimani *et al.*, 2019). To collect and process data, the system utilizes diverse smart devices based on IoT technology (Bekara, 2014; Kimani *et al.*, 2019; Gunduz and Das, 2020). Such strong dependencies on digital communication technology make the smart grids more vulnerable to cyber attacks (Gunduz and Das, 2020; Moghadam *et al.*, 2020). The data collected and processed from smart devices can critically impact people's lives because our lives are closely associated with city facilities and services consisting of a great number of electrical devices. Cyber attacks on these data can lead to breaks in national security, disruption of public order, and loss of life or large-scale economic damage (Vitunskaite *et al.*, 2019; Gunduz and Das, 2020). Consequently, the cyber security risks of the smart grid can be considered as risks to the physical world.

7.2.4 Existing Cyber-Physical Security Risks and Solutions for Smart Grids

The function of smart grids depends significantly on IoT technologies, which make the grids vulnerable to cyber attacks (Gunduz and Das, 2020; Moghadam *et al.*, 2020). This is because the components of smart grids (e.g. smart meters and smart appliances), which are connected with one another, allow cyber-attackers to easily achieve unpermitted access to the smart grid network (Mo *et al.*, 2011). Cyber attacks against smart grids can lead to security, economic, and even safety problems because grids are closely associated with every aspect of people's lives as they reliably, securely, and economically transmit electricity throughout the city (Vitunskaite *et al.*, 2019; Gunduz and Das, 2020). Because of the vulnerability of smart grids and the way they link people's lives, many studies address the need for improved cyber security for smart grids.

Mo *et al.* (2011) pointed out that not only cyber approaches but also physical approaches are required for improving smart grid security which can be compromised by cyber attacks and physical attacks (e.g. using compromised sensors, a shunt to bypass sensors, etc.). Physical attacks can be mitigated through *system theory*, which addresses the properties of the physical system, such as performance, stability, and safety (Mo *et al.*, 2011). They classify the attacks on cyber-physical systems in smart grids according to both cyber and physical aspects. Based on these classifications, they proposed a system-theoretic approach by considering the physical aspects in more detail than traditional security and cryptographic approaches, which traditionally focus on cyber attacks. The authors developed detection algorithms and countermeasures to prevent physical attacks against smart grids. These algorithms and countermeasures can be used to complement the traditional cyber-security approaches for an additional layer of protection. This article verified that combining cyber security and system theory can effectively mitigate the cyber-physical risks in smart grids.

Liu *et al.* (2012) systemically analyzed the cyber vulnerabilities of smart grids in two points of view: *cyber security issues* and *privacy issues*. The study classified *cyber security issues* into five topics (i.e. device, networking, dispatching, and management, anomaly detection, others) and provided possible solutions for the problems in each topic through literature reviews. Regarding the *privacy issues,* the article suggested that private information about where the people were and when and what they were doing may be stolen when energy-related data in smart grids are not properly protected. The article also defined what personal information is, investigated privacy concerns, and provided recommendations for addressing the concerns. However, the article is limited to the overview of cyber security and privacy issues and the recommended possible solutions, but did not verify the effectiveness of the solution-applied systems or theories.

Ashok *et al.* (2014) revealed the increased security risks in the technical initiatives supporting smart grids (e.g. Advanced Metering Infrastructure (AMI), Demand Response (DR), Wide-Area Monitoring, Protection and Control Systems (WAMPAC) based on Phasor Measurement Units (PMU), etc.). The article affirmed that AMI and WAMPAC are more vulnerable to cyber attacks because they heavily depend on cyber infrastructure and its data transfer through several communication protocols to utility control centers and consumers. The article focused on the security of the WAMPAC system because cyber attacks on it can easily cause critical damage to people's lives since the attacks on WAMPAC can impact bulk power system reliability, unlike the attacks on AMI. To protect the Cyber-Physical Security of the WAMPAC system from various coordinated cyber attacks, the article proposed a game-theoretic approach. This approach enables modeling dynamic cyber-attack scenarios, which are useful for obtaining appropriate solution strategies to improve security. However, the article didn't verify how much the solution strategies obtained from this approach mitigate the impact of cyber attacks. To demonstrate the effectiveness, case studies or simulations are required to implement and analyze.

Shapsough *et al.* (2015) discuss four cyber security challenges in smart grids. The challenges contain *connectivity, trust, customer's privacy,* and *software vulnerabilities.* The *connectivity* issue concerns the fact that numerous devices are connected with one another in the grids and that cyber attacks on smart grids might lead to significant damage to people's lives. The *trust* issue concerns with the possibility of users intentionally damaging the smart meter to falsify energy data for their benefit. The *customer's privacy* concerns with the user's critical private information in smart grids that could potentially be exploited by criminals. The *software vulnerabilities* concerns with software used in the smart grids system that could potentially be vulnerable to malware and malicious update. To address these challenges, the authors validated existing security solutions by analyzing them from five aspects (i.e. network security, data security, key management, network security protocols, and compliance checks) and recommended a new conceptual security model for smart grids. The proposed model allows data in smart devices and systems to be directly transmitted to the application layer through WiFi or 4G internet without going through multiple devices or networks.

Gunduz and Das (2020) classified the security threats against IoT-based smart grids and investigated the potential security solutions for each threat type. They investigated the existing cyber attacks from the aspects of confidentiality, integrity, and availability (CIA) and network layers. Based on the evaluation, the research discussed and examined network vulnerabilities, attack countermeasures, and security requirements. However, the article is limited to classifying the solutions with the threat types and frame-working the security threats analysis. Although the classification and framework help understand the cyber attacks against smart grids and identify appropriate solutions for them, the limitations of the solutions need to be examined, and a novel approach that can complement the limitation should be proposed.

As another solution for the security issue of smart grids, Moghadam *et al.* (2020) proposed a security-enhanced protocol for communication between substations and a data center in the smart grid network, which is based on hash and private key to overcome the security weakness associated with the International Electrotechnical Commission (IEC) 62351 standard. IEC 62351 is an industry standard for security in automation systems in the power supply system domain (Schlegel *et al.*, 2017). In their study, the enhanced security of the proposed protocol is verified with AVISPA software.

All the afore-mentioned studies agree that lack of security in smart grids negatively impacts people's lives by interrupting the power system's operation. Even though the studies proposed several solutions for improving the security, more future research on data authenticity and immutability in the smart grid network is still necessary.

7.2.5 What is Smart Building?

The definition of smart building is evolved from the preliminary definition focusing on the technological aspects to the definition focusing more on the interrelationships between occupants and building systems (Martins *et al.*, 2012). According to Linder, Vionnet *et al.* (2017), smart buildings are buildings where the owners, operators, and facility managers can utilize building technology systems (e.g. building management system) connected to a variety of sensors, actuators, and networks with IoT technology in order to improve the reliability and performance of building assets. These systems enable automation control of building systems, promote occupant safety, and facilitate operation management (Sinopoli, 2009). The building technology systems monitor and collect the data of occupants, such as energy usage, user location, and behavior pattern and process the data to generate information that can be used to optimize building services. In this setting, the systems can support people's lives in various ways: inhabitants' comfort, energy savings, time-saving, safety, health and care (Batov, 2015). However, these advantages are accompanied with a risk of cyber attacks that can be extended to the physical world (Tankard, 2016). The building technology systems mainly depend on IoT devices, which are installed all over the smart building (Casado-Vara *et al.*, 2020). Given that IoT devices are vulnerable to risks of cyber security attacks (Bertino,

2016; Qian *et al.*, 2018; Hassan 2019; Amanullah *et al.*, 2020; Waraga *et al.*, 2020), the risks can be substantial to people's lives.

7.2.6 Existing Cyber-Physical Security Risks and Solutions for Smart Buildings

The AECO industry is advancing towards the smart building paradigm in which IoT devices and networks are used to improve occupants' comfort, reduce lifecycle costs, and optimize the operation of building systems (Gao and Pishdad-Bozorgi, 2019a; Gao and Pishdad-Bozorgi, 2019b; Gao *et al.*, 2019; Tang *et al.*, 2019). As the building managers are moving away from the older proprietary systems of the past and adopting new data-intensive, comprehensive building automation and control solutions, there is a desire to gather as much data as possible with lower cost sensors – both wired and wireless (O'Brien, 2019). It is envisioned that in future each building will be 'smart' enough to provide a certain amount of data to the city IoT network in real-time, and the city will provide services in return, such as security, emergency assistance, data connection, and automated operation and maintenance (Gao *et al.*, 2019; Pishdad-Bozorgi *et al.*, 2020). Although the advantages of this trend are undeniable, this substantial change from traditional, isolated, single-function building systems to a "system of systems" integration with existing IoT infrastructures –the Internet and innovative automation devices – exposed smart buildings to significant cyber threats. Moreover, smart buildings are not only subject to known ICT system attacks, but also to a new breed of cyber-physical attacks (Siaterlis *et al.*, 2013).

Cyber attacks on smart buildings alone can have many negative impacts that may pose risks to human safety (O'Brien, 2019). Cyber attacks can impact one or multiple smart building cyber security goals, involving confidentiality, authenticity, integrity, authorization, and availability, non-repudiation (Komninos *et al.*, 2014; Radanliev *et al.*, 2020; Sharma, 2020). Cyber attacks can be classified into passive attacks, which attempt to utilize the data housed in building systems without affecting the operation of systems , or active attacks, which attempt to interrupt system operation or alter its resources (Qi *et al.*, 2017).

In cases of passive attacks, the compromised cyber security of a smart building will lead to data leakage, which may result in occupants' behavior being monitored by the malicious party and/or identity theft. For example, if the attackers have access to data regarding the thermostat setting or occupancy history, with enough time-series data, they can use machine learning techniques to identify the occupants' behavior patterns and thus, know when the occupants will be in the building (or a particular room). Another example is the hack at the target retail chain (Wallace, 2013), in which case the remote access privileges of the HVAC system were exploited to gain access to the target's financial systems, and led to a leak of over 40 million people's credit card information (O'Brien, 2019).

Active attacks tend to disrupt the physical processes and jeopardize the safety properties in the physical world more directly (Wang *et al.*, 2017). If a malicious party hacks into the smart building's security system and turns off the security cameras and intrusion alerts, the physical security of the building will also be compromised. Moreover, if a malicious party can take over the access control system, the building will be completely exposed to physical threats. On the other hand, physical security has always been an important part of cyber security – if adversaries can physically access the building system server, equipment, and even some routing devices, the building's cyber security may be in jeopardy. Therefore, in the smart building domain, the cyber security issues and physical security issues are converging into one critical issue that requires extensive research and innovative solutions.

A limited number of research studies have been conducted to improve the Cyber-Physical Security of smart buildings. An effective approach is to enhance the security of building systems; for example, Wang *et al.* (2017) use a security-enhanced, microkernel architecture to ensure the BAS's security in a hostile cyber environment. Another research trend is to create testbeds to simulate smart buildings for cyber-security experimentation (Mekikis *et al.*, 2013; Tong *et al.*, 2014). Such testbeds are capable of 1) enabling users to specify the building-area network topology, communication protocols and appliances, and developing security mechanisms, such as information flow tracking (Tong *et al.*, 2014), and 2) detecting and localizing events (such as water leakages and system failure) in smart buildings (Mekikis *et al.*, 2013).

The Cyber-Physical Security of smart buildings is an innovative, interdisciplinary topic. Many issues have emerged but not been thoroughly investigated yet, such as the existing and potential attack models, the risk assessment criteria, and countermeasures under different circumstances. Two major research gaps exist in this area – first, the deployment of smart building applications requires increased cyber-physical interdependencies, hence security issues in smart buildings cannot be fully addressed only by considering the cyber layer (Qi *et al.* 2017). There is yet a framework or testbed to examine the joint effects of cyber attack and physical attack, considering the smart building systems and devices, as well as the building layout. Second, humans tend to be the weak link of the Cyber-Physical Security of smart buildings. Studies on the human-building interaction from a security perspective are in need.

7.2.7 What is Smart Device?

The afore-mentioned smart built environments (i.e. smart city, smart grid, and smart building) should collect and process data from the physical world and exchange the data across the systems or devices for operation of the smart built environments. This can be achieved by smart devices. Smart devices perceive information from the environment through sensors; process the information automatically without the direct command of the user; and enable the exchange of information among the systems (Silverio-Fernández *et al.*, 2018). Silverio-Fernández *et al.* (2018) have defined the smart device as follows: "*A smart device is a context-aware electronic device capable of performing autonomous computing and connecting to other devices wire or wirelessly for data exchange.*" The data exchange among smart devices is enabled by IoT technology (Miller, 2015). In addition, the IoT technology enables the systems to sense and control objects, which facilitate integration between the physical world and computer-based systems (Bertino, 2016). However, on the other side, the IoT technology is vulnerable to risks of cyber security attacks (Bertino, 2016; Qian *et al.*, 2018; Hassan, 2019; Amanullah *et al.*, 2020; Waraga *et al.*, 2020). Given that IoT technology is one of the main components for smart built environments (Zanella *et al.* 2014; Waraga *et al.*, 2020), the vulnerability can be a serious threat to people's lives in the smart built environments. Therefore, reliable and foolproof solutions need to be developed by creating new technologies or combining existing technologies to address the security issues (Amanullah *et al.*, 2020).

7.2.8 Existing Cyber-Physical Security Risks and Solutions for Smart Devices

The security requirements of IoT devices involve: 1) identifying the device itself and its administrative entities, such as a gateway, 2) protecting device hijacking, and 3) protecting the information flow between devices and their administrative entities (Minoli *et al.*, 2017; Radanliev *et al.*, 2020). As new IoT devices are being invented and implemented for smart building solutions, these new technologies coexist with legacy building systems, such as HVAC, energy management systems, lighting control systems, video surveillance systems, access control systems, elevator control systems that must be managed, maintained, and gradually modernized. The coexistence of new IoT devices and legacy devices creates a mixed-criticality environment in which new attack vectors are introduced (Wang *et al.*, 2017).

A study showed how an attacker can obtain full control over some smart building devices by injecting arbitrary audio signals into their microphones via light commands (laser) (Sugawara *et al.*). The researchers breached the security measures of some popular voice assistants, such as Amazon Alexa, Apple Siri, Facebook Portal, and Google Assistant, and showed that user authentication of these devices is often lacking or non-existent, allowing the attacker to unlock the target's smart lock-protected front doors, open garage doors, shop on e-commerce websites at the target's expense, or even locate, unlock, and start various vehicles (e.g. Tesla and Ford) that are connected to the target's account. The study revealed several vulnerabilities in today's smart building devices besides highlighting lack of an efficient approach to measure and mitigate the impact of cyber threats on both the cyber and the physical parts of smart buildings.

The cyber-security challenge has always been an obstacle to the popularization of IoT technologies in smart buildings. Minoli *et al.* (2017) summarized the IoT-related challenges in the building sector. Those related to the cyber risks of devices are:

Intrinsic IoT security issues: Most of the software codes in the IoT ecosystems have exploitable vulnerabilities. Moreover, if a device is not protected physically, adversaries can access the device via physical attacks, and the rest of the IoT ecosystem.

Low-complexity devices: Limit the amount of computing power needed for encryption, firewalling, and deep packet analysis.

Limited on-board power: Limits the amount of computing needed for security algorithms.

Accessibility: Devices may be in an open environment, where they can be physically tampered with or stolen.

Device mobility: Portable devices may be placed in some 'foreign' network of unknown security status.

Active system: IoT devices are always connected; hence they are more susceptible to cybersecurity attacks.

System size: Scalable solutions are in need to incorporate a large number of IoT devices in the 'system'.

Many devices and access points are required in smart building deployments, and this presents additional cyber risks because of the devices operating outside the traditional facilities management domain. To solve this problem, enhanced attack prevention, detection, and mitigation approaches should be implemented at both the cyber and physical layers (Qi, Kim *et al.*, 2017). Threats and risks of cyber attacks can be effectively mitigated with the blockchain technology.

Blockchain can generate a data platform for IoT systems in which data theft and tampering are basically impossible with its hash function, consensus mechanism, and capability of distributing the data transaction records to every participant (Samaniego and Deters, 2016). These benefits enable the smart devices to provide additional security to people living in smart built environments.

7.3   Blockchain Technology for Cyber-Physical Security of Smart Built Environments

This section discusses what the blockchain technology (blockchain) is and how it can enhance Cyber-Physical Security of smart built environments. The section provides an explanation of the two main features of blockchain – decentralized network and tamper-proof digital storage, and examines how these two features can enhance the security. It also reviews various studies on blockchain-based security enhancement of the smart built environments.

7.3.1   What is Blockchain Technology and How It can Enhance Cyber-Physical Security on Smart Built Environments?

Blockchain is a database that enables peer-to-peer data transactions in a tamper-proof environment through two main features – *decentralized Network with timestamp* and *linked data list through hash function in conjunction with consensus protocol.*

7.3.2   Decentralized Network with Timestamp

Unlike traditional databases, blockchain doesn't have a single entity that manages and stores data transaction records (Abeyratne and Monfared, 2016; Aung and Tantidham, 2017). Instead, in the blockchain network, all the nodes in the network will be the actual users, whose records blockchain replicates with timestamps of each transaction and then stores them in every node in the network (Fig. 7.2). All the users of this network can see every transaction record with the timestamp. This is the reason why blockchain is called a decentralized (or distributed) ledger. In this ledger, no individual record can be removed unless all the records in all the nodes are removed (Aung and Tantidham, 2017). This prevents the two biggest weaknesses of traditional database: data loss (Jiang *et al.*, 2017; Gatteschi *et al.*, 2018; Saraf and Sabadra, 2018) and single point of failure (Wang *et al.*, 2018; Xiong *et al.*, 2018; Yakubov *et al.*, 2018).

7.3.3   Linked Blocks with Hash and Consensus Protocol

Despite the advantages of decentralization, it cannot ensure complete data integrity and authenticity (Oh, 2017). To achieve data integrity and authenticity, blockchain leverages a *hash function* in conjunction with a *consensus protocol.* The *hash function* is used to convert data into a hash value, which is a random combination of numbers and letters. Any small change in data will give the data a completely different hash value. This feature ensures data integrity and authenticity. For instance, users will see when the existing data have been tampered with (integrity destruction)

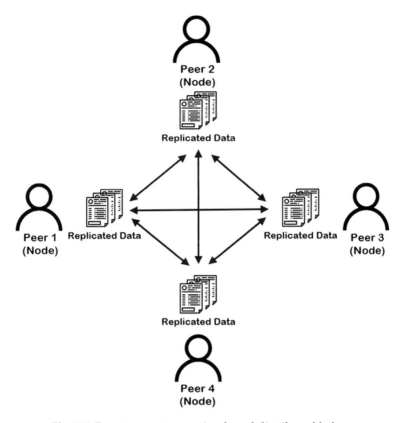

Fig. 7.2: Peer-to-peer transaction-based distributed ledger

or the counterfeited data are newly added (authenticity destruction), as the hash value of the block containing the data will change. The users can verify the integrity and authenticity of the block by comparing the original hash value with the changed hash value. In blockchain, the hash values link blocks with each other. Each block has its own hash value, which is computed of the hash value of the transaction data in the current block, the hash value of the previous block, timestamp, and nonce, which is an arbitrary number. This creates the link between blocks (Fig. 7.3), which makes it much more difficult for anyone to tamper with the transaction data in the blocks (Lisk, 2019).

If an attacker changes any of the data in one block, the attacker must also create a corresponding change in the next block, which includes the changed hash value of the attacked block. However, to create the next new block, the attacker must satisfy the consensus protocol, such as proof-of-work (Mougayar, 2016). The proof-of-work is a process to find out the proper nonce for creating the hash value of the new block (Whittle, 2018).

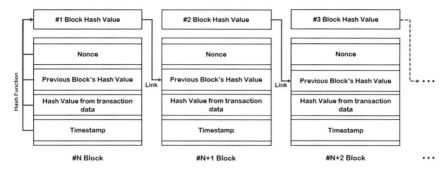

Fig. 7.3: Linked blocks with hash

Even though the attacker can create the new blocks by identifying the proper nonce, the falsified blockchain should be longer than the original blockchain in order to be considered an official blockchain (Fig. 7.4). Because the proof-of-work needs substantial computing power and processing time, it is nearly impossible that the attacker gets falsified blockchain to be the official one by being longer than the original blockchain in which other nodes are continually creating authenticated blocks. In conclusion, in the blockchain network, either tampering with the existing data or adding unauthenticated data is nearly impossible.

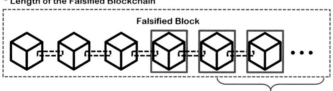

Fig. 7.4: Selecting process for an official blockchain

7.3.4 Enhancing Cyber Security and Privacy of Smart Built Environments through Blockchain

As discussed in previous sections, smart built environments are very vulnerable to cyber attacks, which can lead to significant risks to people's lives. A major reason for vulnerability stems from the two main characteristics of the environment: 1) *cloud database utilization* and 2) *IoT-*

and ICT-based devices application. The cloud database employment causes risks, such as a single point of failure, data loss and breach, and data integrity and authenticity destruction (Xie *et al.*, 2020). The IoT- and ICT-based devices' application causes risks, such as data loss and breach, data integrity and authenticity destruction, and physical attacks on devices. While these risks are the focus of a growing debate about cyber security of smart built environments, a preventive way to eliminate these Cyber-Physical Security risks is often neglected.

The majority of these risks can be eliminated or substantially mitigated by two aforementioned features of blockchain (i.e. *decentralized network* and *linked blocks by hash and consensus protocol*), and this will enhance the Cyber-Physical Security of smart built environments.

The data decentralization of blockchain can effectively prevent the single point of failure and data loss and breach (Xie *et al.*, 2020). Although decentralization is not proper for preventing the data breach because all the data in blockchain should be replicated and shared with every user, this drawback can be complemented by establishing private blockchain with data encryption technology (e.g. the hash function). Furthermore, data links by hash and consensus protocol of blockchain can strongly protect data integrity and authenticity. This can help prevent the destruction of data integrity and authenticity in the operating systems of smart built environments. Despite these two countermeasures from blockchain applications, the smart built environments still face the risk of physical attack on the devices which are used for operating the environment. These findings are summarized in Fig. 7.5.

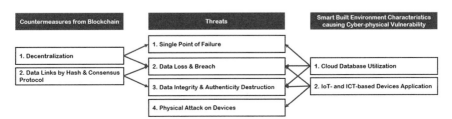

Fig. 7.5: Threats and countermeasures of cyber-physical risks in smart built environments

7.3.5 Literature Reviews on Blockchain Application to Smart Built Environments

Multiple studies have found that blockchain applications effectively address cyber security threats in smart built environments (Biswas and Muthukkumarasamy, 2016; Aung and Tantidham, 2017; Dorri *et al.*, 2017; Mylrea and Gourisetti, 2017; Pop *et al.*, 2018; Qu *et al.*, 2018; Minoli, 2019;

Agung and Handayani, 2020; Makhdoom *et al.*, 2020, Qashlan, Nanda *et al.*, 2020).

Biswas and Muthukkumarasamy (2016) and Makhdoom *et al.* (2020) pointed out that the data collected and processed by IoT systems in smart cities are vulnerable to availability, integrity, and privacy threats. As a countermeasure to the threats, these studies proposed a blockchain-based security framework, which enables users to engage in privacy-preserving and secure IoT data sharing in smart city environments. Other studies (Mylrea and Gourisetti, 2017; Pop *et al.*, 2018; Agung and Handayani, 2020) utilized *smart contract*, which can be implemented by blockchain, to build a more secure and reliable smart grid system against security threats. A *smart contract* is a computer protocol intended to digitally facilitate or enforce automatic data-transaction processes by providing rules and penalties (Rosic, 2016). With a *smart contract*, the data satisfying the rules for transaction can be automatically processed in the system. This can help smart grids to automatically balance energy demand with energy production (Pop, Cioara *et al.*, 2018). The reason why this protocol can function is that the transaction data in blockchain are trustworthy since the blockchain guarantees the integrity and authenticity of the data. In smart grids, blockchain can keep energy transaction data secured and simultaneously can supply reliable electricity more efficiently (Mylrea and Gourisetti, 2017; Pop *et al.*, 2018; Agung and Handayani, 2020). Blockchain application also has the potential to improve the security of smart homes that use IoT devices. Examples can be found in several studies (Aung and Tantidham, 2017; Dorri *et al.*, 2017; Minoli, 2019; Qashlan *et al.*, 2020). Aung and Tantidham (2017) found that private blockchain can improve security and protect the privacy of the IoT device-based data against cyber-attacks. Dorri *et al.* (2017) proposed lightweight instantiation of a blockchain application system for IoT devices in smart homes and verified that the overhead incurred by the proposed system is insignificant when compared with the benefits of security and privacy. Qashlan *et al.* (2020) developed a private blockchain-based smart security solution for smart homes in which only the home owner could access and monitor home appliances, by utilizing ethereum smart contract. The study designed simple smart contracts to enable smart devices to communicate without the need for trusted third party. Qu *et al.* (2018) noted that the existing blockchain is not proper for IoT devices in smart homes, which have less energy and memory. To overcome this limitation, the study proposed a hypergraph-based blockchain model, which can provide more efficient network storage than the traditional blockchain network did, so that the IoT devices can obtain the security and privacy benefits from blockchain.

Even though the benefits of these blockchain applications are understood and accepted, few studies have investigated what type of physical risks can be caused by the security and privacy threats from cyber attacks against smart built environments or how the blockchain applications, which enhance cyber security and privacy, mitigate the physical risks and enhance people's safety in smart built environments. Future research in this area is necessary to investigate if blockchain can be one of the most effective ways to enhance Cyber-Physical Security. Also necessary is future research regarding the assessment parameters and the testbed framework.

7.4 Conclusion

As smart built environments are continually growing and evolving, so do cyber-physical risks. These risks can negatively impact people's lives. Cyber attacks against smart built environments can lead to city service malfunction, energy supply failure, private data breach, building operation system misuse, etc. Despite these significant impacts, most current research has focused on cyber security. Relatively few studies have investigated cyber-physical threats and risks in smart built environments or proposed proper countermeasures for preventing them.

One effective countermeasure can be blockchain, which has the potential to serve as a data platform for IoT systems in which data theft and tampering are basically impossible because of its hash function, consensus mechanism, and capability of distributing the data transaction records to every participant (Samaniego and Deters, 2016). These benefits can enhance cyber security of smart built environments. Multiple studies have demonstrated that blockchain applications effectively address cyber security threats in smart built environments. Nevertheless, the studies on physical risks caused by the cyber attacks are still missing, and few studies have addressed how the cyber security achieved by blockchain can extend to people's lives in smart built environments.

This chapter contributes to the body of knowledge on Cyber-Physical Security by discussing the definition of each layer of smart built environment and reviewing the relevant literature regarding the cyber-physical risks. The literature reviews investigate risks and explore the existing solutions for mitigating the risks. Through these reviews, this chapter demonstrates that more studies on cyber-physical risks and threats are required to develop proper countermeasures. This chapter also includes reviews of existing literature on blockchain for the smart built environments. It demonstrates that blockchain technology can enhance the Cyber-Physical Security in smart built environments through its security-enhancing system. However, more studies involving use cases are required to demonstrate how blockchain enhances cyber security and impacts people's lives.

References

Abeyratne, S.A. and Monfared, R.P. (2016). Blockchain ready manufacturing supply chain using distributed ledger. *International Journal of Research in Engineering and Technology*, 5(9): 1-10.

Agung, A.A.G. and Handayani, R. (2020). Blockchain for smart grid. *Journal of King Saud University - Computer and Information Sciences*. (In Press)

AlDairi, A. (2017). Cyber security attacks on smart cities and associated mobile technologies. *Procedia Computer Science*, 109: 1086-1091.

Amanullah, M.A., Habeeb, R.A.A., Nasaruddin, F.H., Gani, A., Ahmed, E., Nainar, A.S.M., Akim, N.M. and Imran, M. (2020). Deep learning and big data technologies for IoT security. *Computer Communications*, 151: 495-517.

Ashok, A., Hahn, A. and Govindarasu, M. (2014). Cyber-physical security of wide-area monitoring, protection and control in a smart grid environment. *Journal of Advanced Research*, 5(4): 481-489.

Aung, Y.N. and Tantidham, T. (2017). Review of Ethereum: Smart home case study. 2017 2nd International Conference on Information Technology (INCIT), IEEE.

Baig, Z.A., Szewczyk, P., Valli, C., Rabadia, P., Hannay, P., Chernyshev, M., Johnston, M., Kerai, P., Ibrahim, A. and Sansurooah, K. (2017). Future challenges for smart cities: Cyber-security and digital forensics. *Digital Investigation*, 22: 3-13.

Batov, E.I. (2015). The distinctive features of 'smart' buildings. *Procedia Engineering*, 111: 103-107.

Bekara, C. (2014). *Security Issues and Challenges for the IoT-based Smart Grid*. FNC/MobiSPC.

Bertino, E. (2016). *Data Security and Privacy in the IoT*. EDBT.

Biswas, K. and Muthukkumarasamy, V. (2016). Securing smart cities using blockchain technology. 2016 IEEE 18th International Conference on High Performance Computing and Communications; IEEE 14th International Conference on Smart City; IEEE 2nd International Conference on Data Science and Systems (HPCC/SmartCity/DSS), IEEE.

Casado-Vara, R., Martin-del Rey, A., Affes, S., Prieto, J. and Corchado, J.M. (2020). IoT network slicing on virtual layers of homogeneous data for improved algorithm operation in smart buildings. *Future Generation Computer Systems*, 102: 965-977.

Chamoso, P., González-Briones, A., De La Prieta, F., Venyagamoorthy, G.K. and Corchado, J.M. (2020). Smart city as a distributed platform: Toward a system for citizen-oriented management. *Computer Communications*, 152: 323-332.

Chatterjee, S., Kar, A.K., Dwivedi, Y.K. and Kizgin, H. (2018). Prevention of cyber crimes in smart cities of India: From a citizen's perspective. *Information Technology & People*, 32(5): 1153-1183.

Chauhan, S., Agarwal, N. and Kar, A.K. (2016). Addressing big data challenges in smart cities: A systematic literature review. *Info*, 18(4): 73-90.

Chung, H., Park, J., Lee, S. and Kang, C. (2012). Digital forensic investigation of cloud storage services. *Digital Investigation*, 9(2): 81-95.

Delgado-Gomes, V., Martins, J.F., Lima, C. and Borza, P.N. (2015). Smart grid security issues. 2015 9th International Conference on Compatibility and Power Electronics (CPE), IEEE.

Dileep, G. (2020). A survey on smart grid technologies and applications. *Renew. Energy*, 146: 2589-2625.

Do, Q., Martini, B. and Choo, K.K.R. (2018). Cyber-physical systems information gathering: A smart home case study. *Computer Networks*, 138: 1-12.

Dorri, A., Kanhere, S.S., Jurdak, R. and Gauravaram, P. (2017). Blockchain for IoT security and privacy: The case study of a smart home. 2017 IEEE International Conference on Pervasive Computing and Communications Workshops (PerCom workshops), IEEE.

Gao, X. and Pishdad-Bozorgi, P. (2019). BIM-enabled facilities operation and maintenance: A review. *Advanced Engineering Informatics*, 39: 227-247.

Gao, X. and Pishdad-Bozorgi, P. (2019). A framework of developing machine learning models for facility lifecycle cost analysis. *Building Research & Information*, 1-25.

Gao, X., Pishdad-Bozorgi, P., Shelden, D. and Tang, S. (2019). A Scalable Cyber-Physical System Data Acquisition Framework for the Smart Built Environment. The 2019 ASCE International Conference on Computing in Civil Engineering, Atlanta, GA, ASCE.

Gatteschi, V., Lamberti, F., Demartini, C., Pranteda, C. and Santamaria, V. (2018). To blockchain or not to blockchain: That is the question. *IT Professional*, 20(2): 62-74.

Gračanin, D., Matković, K. and Wheeler, J. (2015). An approach to modeling internet of things based smart built environments. 2015 Winter Simulation Conference (WSC), IEEE.

Gunduz, M.Z. and Das, R. (2020). Cyber-security on smart grid: Threats and potential solutions. *Computer Networks*: 107094.

Harrison, C., Eckman, B., Hamilton, R., Hartswick, P., Kalagnanam, J., Paraszczak, J. and Williams, P. (2010). Foundations for smarter cities. *IBM Journal of Research and Development*, 54(4): 1-16.

Hasbini, M.A., Eldabi, T. and Aldallal, A. (2018). Investigating the information security management role in smart city organisations. *World Journal of Entrepreneurship, Management and Sustainable Development, Info*, 18(4): 73-90.

Hassan, W.H. (2019). Current research on Internet of Things (IoT) security: A survey. *Computer Networks*, 148: 283-294.

Jia, M., Komeily, A., Wang, Y. and Srinivasan, R.S. (2019). Adopting Internet of Things for the development of smart buildings: A review of enabling technologies and applications. *Automation in Construction*, 101: 111-126.

Jiang, P., Guo, F., Liang, K., Lai, J. and Wen, Q. (2017). Searchain: Blockchain-based private keyword search in decentralized storage. *Future Generation Computer Systems*, 107: 787-792.

Kimani, K., Oduol, V. and Langat, K. (2019). Cyber security challenges for IoT-based smart grid networks. *International Journal of Critical Infrastructure Protection*, 25: 36-49.

Komninos, N., Philippou, E. and Pitsillides, A. (2014). Survey in smart grid and smart home security: Issues, challenges and countermeasures. *IEEE Communications Surveys & Tutorials*, 16(4): 1933-1954.

Laufs, J., Borrion, H. and Bradford, B. (2020). Security and the smart city: A systematic review. *Sustainable Cities and Society*, 102023.

Linder, L., Vionnet, D., Bacher, J.-P. and Hennebert, J. (2017). Big Building Data – A Big Data platform for smart buildings. *Energy Procedia*, 122: 589-594.

Lisk. (2019). What is Blockchain? https://lisk.io/what-is-blockchain

Liu, J., Xiao, Y., Li, S., Liang, W. and Chen, C.P. (2012). Cyber security and privacy issues in smart grids. *IEEE Communications Surveys and Tutorials*, 14(4): 981-997.

Lombardi, P., Giordano, S., Farouh, H. and Yousef, W. (2012). Modelling the smart city performance. *Innovation: The European Journal of Social Science Research*, 25(2): 137-149.

Makhdoom, I., Zhou, I., Abolhasan, M., Lipman, J. and Ni, W. (2020). Privy Sharing: A blockchain-based framework for privacy-preserving and secure data sharing in smart cities. *Computers and Security*, 88: 101653.

Marsal-Llacuna, M.-L., Colomer-Llinàs, J. and Meléndez-Frigola, J. (2015). Lessons in urban monitoring taken from sustainable and livable cities to better address the Smart Cities initiative. *Technological Forecasting and Social Change*, 90: 611-622.

Martins, J., Oliveira-Lima, J., Delgado-Gomes, V., Lopes, R., Silva, D., Vieira, S. and Lima, C. (2012). Smart homes and smart buildings. 2012 13th Biennial Baltic Electronics Conference, IEEE.

Mekikis, P.-V., Athanasiou, G. and Fischione, C. (2013). A wireless sensor network testbed for event detection in smart homes. 2013 IEEE International Conference on Distributed Computing in Sensor Systems, IEEE.

Miller, M. (2015). *The Internet of Things: How Smart TVs, Smart Cars, Smart Homes, and Smart Cities are Changing the World*. Pearson Education.

Minoli, D. (2019). Positioning of blockchain mechanisms in IOT-powered smart home systems: A gateway-based approach. *Internet of Things*, 100147.

Minoli, D., Sohraby, K. and Occhiogrosso, B. (2017). IoT considerations, requirements, and architectures for smart buildings – Energy optimization and next-generation building management systems. *IEEE Internet of Things Journal*, 4(1): 269-283.

Mo, Y., Kim, T.H.-J., Brancik, K., Dickinson, D., Lee, H., Perrig, A. and Sinopoli, B. (2011). Cyber-physical security of a smart grid infrastructure. Proceedings of the IEEE, 100(1): 195-209.

Moghadam, M.F., Nikooghadam, M., Mohajerzadeh, A.H. and Movali, B. (2020). A lightweight key management protocol for secure communication in smart grids. *Electric Power Systems Research*, 178: 106024.

Mohanty, S.P., Choppali, U. and Kougianos, E. (2016). Everything you wanted to know about smart cities: The internet of things is the backbone. *IEEE Consumer Electronics Magazine*, 5(3): 60-70.

Monostori, L. (2018). Cyber-physical systems. The International Academy for Production. *CIRP Encyclopedia of Production Engineering*: 1-7.

Mougayar, W. (2016). *The Business Blockchain: Promise, Practice, and Application of the Next Internet Technology*. John Wiley & Sons.

Mylrea, M. and Gourisetti, S.N.G. (2017). *Blockchain for Smart Grid Resilience: Exchanging Distributed Energy at Speed, Scale and Security*, 2017 Resilience Week (RWS), IEEE.

Nam, T. and Pardo, T.A. (2011). Conceptualizing smart city with dimensions of technology, people, and institutions. Proceedings of the 12th Annual International Digital Government Research Conference: Digital Government Innovation in Challenging Times.

O'Brien, L. (2019). Cybersecurity for Smart Buildings. https://www.arcweb.com/blog/cybersecurity-smart-buildings.

Oh, M.W. (2017). "블록체인한번에이해하기 (Korean Title). https://homoefficio. github.io/2017/11/19/%EB%B8%94%EB%A1%9D%EC%B2%B4%EC%9D %B8-%ED%95%9C-%EB%B2%88%EC%97%90-%EC%9D%B4%ED%95%B4% ED%95%98%EA%B8%B0/

Overman, T.M., Sackman, R.W., Davis, T.L. and Cohen, B.S. (2011). High-assurance smart grid: A three-part model for smart grid control systems. *Proceedings of the IEEE*, 99(6): 1046-1062.

Paridari, K., Mady, A.E.-D., La Porta, S., Chabukswar, R., Blanco, J., Teixeira, A., Sandberg, H. and Boubekeur, M. (2016). Cyber-Physical-Security framework for building energy management system. 2016 ACM/IEEE 7th International Conference on Cyber-Physical Systems (ICCPS), IEEE.

Pishdad-Bozorgi, P., Shelden, D. and Gao, X. (2020). Introduction to Cyber-Physical Systems in the Built Environment. Sawhney, Riley, M. and Irizarry, J. (Ed.). Construction 4.0: An Innovation Platform for the Built Environment. Taylor and Francis.

Pop, C., Cioara, T., Antal, M., Anghel, I., Salomie, I. and Bertoncini, M. (2018). Blockchain based decentralized management of demand response programs in smart energy grids. *Sensors*, 18(1): 162.

Qashlan, A., Nanda, P. and He, X. (2020). Automated ethereum smart contract for block chain based smart home security. pp. 313-326. *In:* Smart Systems and IoT: Innovations in Computing. Springer.

Qi, J.J., Kim, Y., Chen, C., Lu, X.N. and Wang, J.H. (2017). Demand response and smart buildings: A survey of control, communication, and cyber-physical security. *ACM Transactions on Cyber-Physical Systems*, 1(4): 1-25.

Qian, Y., Jiang, Y., Chen, J., Zhang, Y., Song, J., Zhou, M. and Pustišek, M. (2018). Towards decentralized IoT security enhancement: A blockchain approach. *Computers & Electrical Engineering*, 72: 266-273.

Qu, C., Tao, M. and Yuan, R. (2018). A hypergraph-based blockchain model and application in Internet of Things-enabled smart homes. *Sensors*, 18(9): 2784.

Radanliev, P., De Roure, D.C., Nurse, J.R., Montalvo, R.M., Cannady, S., Santos, O., Burnap, P. and Maple, C. (2020). Future developments in standardisation of cyber risk in the Internet of Things (IoT). *SN Applied Sciences*, 2(2): 169.

Raut, M.M., Sable, R.R. and Toraskar, S.R. (2016). Internet of Things (IoT) based smart grid. *International Journal of Engineering Trends and Technology (IJETT)*, 34: 15-20.

Rosic, A. (2016). Smart Contracts: The blockchain technology that will replace lawyers. Retrieved from https://blockgeeks.com/guides/smart-contracts/

Samaniego, M. and Deters, R. (2016). Blockchain as a Service for IoT. 2016 IEEE International Conference on Internet of Things (iThings) and IEEE Green Computing and Communications (GreenCom) and IEEE Cyber, Physical and Social Computing (CPSCom) and IEEE Smart Data (Smart Data), IEEE.

Saraf, C. and Sabadra, S. (2018). Blockchain platforms: A compendium. 2018 IEEE International Conference on Innovative Research and Development (ICIRD), IEEE.

Schlegel, R., Obermeier, S. and Schneider, J. (2017). A security evaluation of IEC 62351. *Journal of Information Security and Applications*, 34: 197-204.

Shapsough, S., Qatan, F., Aburukba, R., Aloul, F. and Al Ali, A. (2015). Smart grid cyber security: Challenges and solutions. 2015 International Conference on Smart Grid and Clean Energy Technologies (ICSGCE), IEEE.

Sharma, A. (2020). Impact of cyber risks from Internet of Things. *TEST Engineering & Management*, 82: 2484-2487.

Siaterlis, C., Genge, B. and Hohenadel, M. (2013). EPIC: A testbed for scientifically rigorous cyber-physical security experimentation. *IEEE Transactions on Emerging Topics in Computing*, 1(2): 319-330.

Silverio-Fernández, M., Renukappa, S. and Suresh, S. (2018). What is a smart device? A conceptualisation within the paradigm of the Internet of Things. *Vizualization in Engineering*, 6(1): 3.

Sinopoli, J.M. (2009). *Smart Buildings Systems for Architects, Owners and Builders*. Butterworth-Heinemann.

SmartGrids, E.T.P. (2012). SmartGrids SRA 2035 Strategic Research Agenda Update of the SmartGrids SRA 2007 for the needs by the year 2035. *Smart Grids European Technology Platform*: 74.

Sookhak, M. (2015). *Dynamic Remote Data Auditing for Securing Big Data Storage in Cloud Computing*. University of Malaya.

Sugawara, T., Cyr, B., Rampazzi, S., Genkin, D. and Fu, K. (2020). Light commands: Laser-based audio injection attacks on voice-controllable systems. *In*: 29th {USENIX} Security Symposium ({USENIX} Security 20), 2631-2648.

Tang, S., Shelden, D., Eastman, C., Pishdad-Bozorgi, P. and Gao, X. (2019). A review of Building Information Modeling (BIM) and Internet of Things (IoT) devices integration: Present status and future trends. *Automation in Construction*, 101: 127-139.

Tankard, C. (2016). Smart buildings need joined-up security. *Network Security*: 1.

Tong, J., Sun, W. and Wang, L. (2014). A smart home network simulation testbed for cybersecurity experimentation. International Conference on Testbeds and Research Infrastructures. Springer.

Tully, G., Cohen, N., Compton, D., Davies, G., Isbell, R. and Watson, T. (2020). Quality standards for digital forensics: Learning from experience in England & Wales. *Forensic Science International: Digital Investigation*, 200905.

Vaccaro, A., Pepiciello, A. and Zobaa, A.F. (2020). Introductory chapter: Open problems and enabling methodologies for smart grids. Research Trends and Challenges in Smart Grids. IntechOpen.

Vitunskaite, M., He, Y., Brandstetter, T. and Janicke, H. (2019). Smart cities and cyber security: Are we there yet? A comparative study on the role of standards, third party risk management and security ownership. *Computers & Security*, 83: 313-331.

Wallace, G. (2013). *Target Credit Card Hack: What You Need to Know*. https://money.cnn.com/2013/12/22/news/companies/target-credit-card-hack/

Wang, P., Ali, A. and Kelly, W. (2015). Data security and threat modeling for smart city infrastructure. 2015 International Conference on Cyber Security of Smart Cities, Industrial Control System and Communications (SSIC), IEEE.

Wang, S., Zhang, Y. and Zhang, Y. (2018). A blockchain-based framework for data sharing with fine-grained access control in decentralized storage systems. *IEEE Access*, 6: 38437-38450.

Wang, X., Habeeb, R., Ou, X., Amaravadi, S., Hatcliff, J., Mizuno, M., Neilsen, M., Rajagopalan, S.R. and Varadarajan, S. (2017). Enhanced security of building automation systems through microkernel-based controller platforms. 2017

IEEE 37th International Conference on Distributed Computing Systems Workshops (ICDCSW), IEEE.

Wang, X., Mizuno, M., Neilsen, M., Ou, X., Rajagopalan, S.R., Boldwin, W.G. and Phillips, B. (2015). Secure rtos architecture for building automation. Proceedings of the First ACM Workshop on Cyber-Physical Systems-Security and/or Privacy.

Waraga, O.A., Bettayeb, M., Nasir, Q. and Talib, M.A. (2020). Design and implementation of automated IoT security testbed. *Computers & Security*, 88: 101648.

Whittle, B. (2018). *What is a Nonce? A No-Nonsense Dive into Proof of Work.* https://coincentral.com/what-is-a-nonce-proof-of-work/

Xie, S., Zheng, Z., Chen, W., Wu, J., Dai, H.-N. and Imran, M. (2020). Blockchain for cloud exchange: A survey. *Computers & Electrical Engineering*, 81: 106526.

Xiong, Z., Zhang, Y., Niyato, D., Wang, P. and Han, Z. (2018). When mobile blockchain meets edge computing. *IEEE Communications Magazine*, 56(8): 33-39.

Yakubov, A., Shbair, W., Wallbom, A. and Sanda, D. (2018). A blockchain-based pki management framework. The First IEEE/IFIP International Workshop on Managing and Managed by Blockchain (Man2Block) colocated with IEEE/IFIP NOMS, 2018, Tapei, Tawain. 23-27 April 2018.

Zanella, A., Bui, N., Castellani, A., Vangelista, L. and Zorzi, M. (2014). Internet of things for smart cities. *IEEE Internet of Things Journal*, 1(1): 22-32.

Zang, L., Yu, Y., Xue, L., Li, Y., Ding, Y. and Tao, X. (2017). Improved dynamic remote data auditing protocol for smart city security. *Personal and Ubiquitous Computing*, 21(5): 911-921.

Zubizarreta, I., Seravalli, A. and Arrizabalaga, S. (2016). Smart city concept: What it is and what it should be. *Journal of Urban Planning and Development*, 142(1): 04015005.

eLUX: The Case Study of Cognitive Building in the Smart Campus at the University of Brescia

Lavinia Chiara Tagliabue

University of Brescia, DICATAM - Department of Civil, Environmental, Architectural Engineering and Mathematics, Via Branze 43, 25121 Brescia, Italy

8.1 Introduction

The complete accomplishment of energy saving through retrofit actions for existing buildings is related to the users' behavior affecting the realization of the investment. Actually, a well-known phenomenon, the 'performance gap' (Shi *et al.*, 2019; Liang *et al.*, 2019; Cuerda *et al.*, 2020) or the 'rebound effect' (Santos *et al.*, 2018; Seebauer, 2018; Thomas and Rosenow, 2020) can produce a discrepancy in the estimated performance of the building, resulting in an energy depletion which is distant from the zero energy balance. Examining the profile of the use of assets, the connection between the variables – occupancy, internal gains, ventilation, set-point temperatures – is, thus, substantial to recognize thresholds of the uncertainty of the energy performance data. Moreover, it is possible to simulate in a virtual environment the occupancy streams with the aim to arrange for a Pre-Occupancy Evaluation (POE). This procedure can contribute to gain early feedback on the real effect that is possible to accomplish by deploying efficiency strategies. In large public buildings and educational facilities, such as schools and university buildings, energy-saving is strongly contingent to HVAC system optimization.

Usually, the HVAC control relies on static models assuming average occupancy ranges. Shifting the paradigm, a cognitive building approach

Email: lavinia.tagliabue@unibs.it

contemplates an Energy Management System (EMS) which is able to adjust the HVAC system based on an occupancy rate model shaped to embrace users' habits and chasing an adequate level of Indoor Air Quality (IAQ), controlled by real-time IoT sensors and actuators. This process has been applied in the eLUX lab – the cognitive building in the smart campus of the University of Brescia, Italy. In this case study, the first at national level, the IAQ data gathered by the IoT network allows refinement of the results of occupancy percentage models. In this way it is conceivable to improve the indoor comfort conditions in the building, increase the users' learning performance in educational facilities and furthermore promote energy saving and proficient use of the renewable energy required to be integrated to partially fulfill the energy needs.

The nearly Zero Energy Buildings (nZEBs) goal, encouraged in the last 10 years by EU Commission Directives (EU Directive, 2010, 2012, 2018) in the member States, is boosting important results in refurbishing the European building stock, both through efficiency policies and application of best practices. A general upgrading in the envelope-energy performance in winter and summer period and the replacement of obsolete HVAC systems with high-efficiency solutions provide significant energy effects which are able to guarantee buildings' lifecycle, high performance, and a decrease in energy need and operational costs.

The digital tool of Energy Performance Certification, introduced in 2010 (EU Directive, 2010) to enhance transparency and the possibility for the customers to have a, efficiency-based comparison of the buildings in the real estate market, has allowed also to qualify the assets from the lifecycle-performance viewpoint, by means of a measurable and quantitative indicator. Therefore, this helps to target the definition of quality and accordingly the appeal of the asset from the user's perspective. Nevertheless, this process of making unbiased the energy performance value has proved to be partly superficial and the deterministic approach to the energy consumption calculation has demonstrated low accuracy and overlapping problems when the compliance with actual performance is concerned. The predicted gap has been measured and it is now evident and expected in Architecture, Engineering, Construction, and Operations (AECO) industry (Hijazi, 2015) where BIM-based Augmented Reality (AR) tools are suggested to bridge the gap (Cheng *et al.*, 2020) and the actual occupancy profile has been assumed to shrink the performance gap (Oliver *et al.*, 2019).

As shown, the penetration of innovative technologies in energy saving is high; nonetheless, tangible results on the final consumption decrease are frequently distant from the expected and predictable outcomes where a substantial deviation from the energy targets is measured (De Wilde, 2014). In this sector, the opportunity to perform a POE evaluation (Menezes *et al.*, 2012) or install an IoT network for onsite monitoring campaign is crucial (Cali *et al.*, 2016) for real compliance with the European directives. Regularly an important variance between the predicted energy performance at the design stage, for new buildings or retrofitted assets, and the measured energy consumption in the operational phase throughout the useful life occurs. This turns into an extremely pertinent matter when the goal is a Net Zero Energy Building and when energy self-sufficiency and carbon neutral balance is a declared target sold to the client.

Probabilistic approaches (Re Cecconi *et al.*, 2017; Voss *et al.*, 2010) have been tested to describe the distribution of consumption data through a Probabilistic Distribution Function (PDF). This method overcomes the energy-classification methodology based on a single value label but percentage thresholds delineate the range within the energy performance value which is delimited according to the building conditions. The current approach is sponsored by energy certification standard adopting a user-standard user-based approach followed by the performance gap occurrence. The probabilistic approach defines a set of possible and probable energy data and sets up a confidence threshold for which it is possible to calculate a probabilistic approximation. This can specify, for example, that in 97 per cent of cases the building consumption will be higher than a certain threshold. The probabilistic methodology envisions scenarios in which the Energy Performance Contracts (EPC) could be modulated on a realistic building's energy behavior with noteworthy matching with the actual use.

It is worthy to note how additional analyses on the reasons of the performance gap can be outlined, considering the accuracy of energy models and the methods adopted by energy modelers who have shown extensive variation of ontological and epistemological attitudes (Kampelis *et al.*, 2017, Figueiredo *et al.*, 2018) to the problem of modeling and the variables to be encompassed as central for energy evaluation.

Moreover, statistically significant samples revealed that not a minor share of the discrepancy among simulated and actual performance relies on the human factor not only in the use phase (i.e. occupancy fluctuation and changes between the designed use of the use of space made by users) but formerly in the assumptions of the modeling phase to detect from +18 per cent to -50 per cent variation in the estimation of the heating value by dynamic predictive models compared to the actual value, based on more than 100 experienced energy modelers (Imam *et al.*, 2017).

8.2 The nZEB Goal Among Performance Gap and Rebound Effect

As above mentioned, the subject of the nonconformity among the measured consumption values and the estimated ones by predictive models is substantial, for instance in educational facilities such as university buildings an average performance gap about +68% for heating consumption and about +53 per cent for electricity has been documented although the *range* can be considerable wider up to values of +115-200 per cent (Perez-Lombard *et al.*, 2008; Menezes *et al.*, 2012). This intense deviation can be traced back to a number of influences such as the evaluation algorithms and modelling procedures, the digital tools and the approximations that these techniques automatically hold depending on the different project phase (Johnston *et al.*, 2014).

It is confirmed that even if the design is definite, performance inconsistencies may still occur if the method implemented by the design team doesn't take into account some critical issues, such as buildability, simplicity, or construction arrangement of the building portions and systems (De Wilde, 2014). Other concerns are linked to the specificity and complexity of advanced systems and technologies, and their control systems setup. As far as the operation phase is concerned, it is obvious that the discrepancies that arise in the implementation phase among the executive design and the as-built description can broadly change the real performance. Furthermore, in the use phase, climate and occupancy variables involve substantial effects that can give high values of performance gap (Zero Carbon Hub, 2014). For example, not all the time the prediction on the use of a building can be completely accurate if only a forecast statement is delivered; occasionally, a refurbishment intervention can enrich the building and the new purposes and uses can have unpredictable consequences on the region, making the building a new hotspot and a magnetic center for the community – on one hand exceeding the estimation defined in the design phase (Bosia 2013; Butera, 2013), on the other, renovating urban areas in depressed contexts (Vicente *et al.*, 2015; Yau *et al.*, 2008).

The rate of building renovation in the EU28 accounts for 1.5 per cent per year; nevertheless, retrofitting interventions are not congruently detached from the performance gap and an emergent anxiety about intensified consumption of energy services is consequential to energy efficiency progress. The effect is named as 'rebound effect' or 'energy saving deficit' or 'energy performance gap' (Yilmaz, 2019; Haas and Biermayr, 2000).

Specific studies on this topic (Galvin, 2014; Lu *et al.*, 2017) establish that the expected savings with energy-saving actions and an upgrading of the energy-service efficacy do not every time produce the expected outcomes. Therefore the estimation of foreseen performance matched to the actual value after renovation becomes a concern of conflict, bearing in mind the impact of the different users' types (Corrado *et al.*, 2016).

For the above-mentioned reasons, a monitoring campaign to double check the energy consumption can solve this problem and allow information on the building usage. The collected data turns into insight for deep analyses on performance gaps and suitable procedures to suggest and confirm the expected outcomes.

The potential application of renovation actions, aimed at shrinking the heating demand by improving thermal insulation, is demonstrated. Nevertheless, the actual performance may turn out to be conflicting from the estimated value because of manifold reasons – installation unconformities, conductivity variation of the insulation materials as a consequence of the fluctuation of hygro-thermal conditions, infiltrations and condensation that may arise throughout the usage, extensive complications in the exact installation procedures of multi-layered insulation systems, for example, in the case of External Thermal Insulation Composite Systems (ETICS). Additionally, users' behaviors as they may not be aware about how to set up the domotic system or the Building Automated System (BAS) or mostly how to achieve energy saving, compromise considerably the actual energy reduction that might be accomplished in comparison with those intended by predictive models with standard occupancy profile (Ioannidis *et al.*, 2017; Yang, 2014).

It is conceivable to assume that in all buildings these complications practically happen since, consistently, when the internal conditions in highly automated buildings do not encounter the comfort level expected by the users, every time (if possible) they act to preserve their specific comfort level by non-optimized setting rather than acclimatizing to an automated system that does not permit corrective arrangements made by the customer.

Therefore, the BAS or application of a manual switch within a BAS, is executed by users in order to modify the internal conditions. This is tough to crack and should not be disregarded where the project of an automated control system to chase energy saving in the building is concerned.

8.2.1 Rebound Effect in Housing Buildings

An important research on the 'rebound effect' (Corrado *et al.*, 2016) calculated the effect of occupant behavior on energy consumption after an asset retrofit. The energy performance of a residential building is calculated after the application of the same energy-efficiency action by considering the user behavior variable, both before and after the restoration, when the dweller routines are influenced by the rebound effect. The rebound effect RB is the variation among the predicted energy savings EP and the effective post-retrofit energy decrement (EP_{RB}), including the change of the dweller behavior after a renovation intervention, as defined in equation (1):

$$RB = \Delta EP - \Delta EP_{RB} = EP_{EB} - EP_{REF} - (EP_{EB} - EP_{REF,RB}) = EP_{REF,RB} - EP_{REF} \quad (1)$$

Starting from Galvin's investigation (Galvin, 2014), the following factors related to the rebound effect and expected energy savings were described:

$I_{RB,1}$ is the amount of rebound effect as a percentage of the energy performance after renovation, with no dweller-behavior variations, as described in the following formula (2):

$$I_{RB,1} = \frac{RB}{EP_{REF}} = \frac{EP_{REF,RB} - EP_{REF}}{EP_{REF}} \tag{2}$$

$I_{RB,2}$ is the fraction of predictable energy savings, which are not a consequence of the rebound effect, as expressed in the following formula (3):

$$I_{RB,2} = \frac{\Delta EP_{RB}}{\Delta EP} = \frac{EP_{EB} - EP_{REF,RB}}{EP_{EB} - EP_{REF}} \tag{3}$$

The assessments (Corrado *et al.*, 2016) for three types of dwellers and housing, in the Italian climate zone E (D.P.R. 412/93) and subjected to energy retrofit are summarized in Table 8.1.

Table 8.1: Variables Adopted for Comparison and Calculation

n	*Dweller*		*Building typology*		*Energy-saving measure*
U1	Elderly single	a)	Apartment blocks	A)	Thermal insulation of the envelope
U2	Young couple with children	b)	Multi-family house	B)	Solar shading devices installation
U3	Adult couple with teenagers	c)	Single-family house	C)	HVAC substitution and control system
				D)	Total refurbishment towards NZEB

Research reveals that major renovations tending to grasp the nearly zero energy building target, gain the lowest level of non-renewable primary energy. However the envelope improved thermal insulation is a very effective retrofit strategy for existing buildings which lack envelope resistance and therefore show a poor performance (i.e. old buildings with no insulation coating or with low insulation). In the following figures, the results are plotted. In Fig. 8.1 the variation between the estimated energy saving and the actual one is described, while in Fig. 8.2, the actual energy saving is computed.

Counting the rebound effect, the net energy demand for heating for all dwellers, analyzed after the envelope improvement and the transformation towards nZEB, drops by 70-75 per cent. Furthermore, energy savings

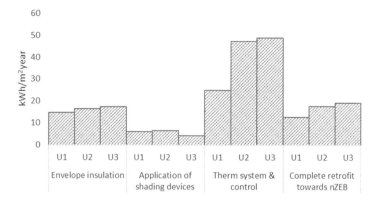

Fig. 8.1: Rebound Effect (RB)

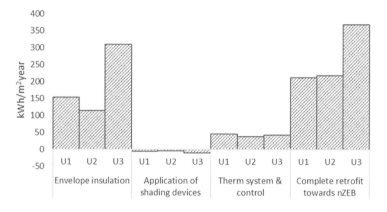

Fig. 8.2: Actual post-retrofit savings($\otimes EP_{RB}$)

would be 5-10 per cent higher without counting the dweller's behavioral change after the renovation. Thermal system replacement for heating and domestic hot water increases the net electricity consumption by 10-30 per cent because of the new control system and the set-point temperature and air changes embraced by the dwellers.

The rebound effect on cooling results in divergence between the situation recorded for the different dwellers and the energy retrofit strategies applied. For U1 and U2. The envelope improvement exacerbates the cooling need by 25-40 per cent%; for U3, the same strategy saves around 10 per cent of energy for cooling. The net cooling demand drop would be significant, with a range between 25-30 per cent, using the solar shading devices. However, the dweller behavioral changes nearly frustrate totally this benefit. The total refurbishment decreases the cooling demand for all the dwellers, but mainly for U2 and U3 by 25-45 per cent.

The total non-renewable primary energy shows decrement for all the dwellers after the envelope insulation (45-70 per cent) and the upgrading towards nZEB (80-85 per cent), even though the saving is narrow if the measure is vertical on the thermal system (10-20 per cent). On the other hand, after the application of solar shading devices, the total non-renewable primary energy marginally increases (2 per cent) and with regard to energy services, non-renewable primary energy for heating results as most sensitive among the analyzed renovation strategies. A higher value of $I_{RB,1}$ emphasizes the amount of energy saving which occurs since a major renovation drives the building to a nZEB result (Fig. 8.3).

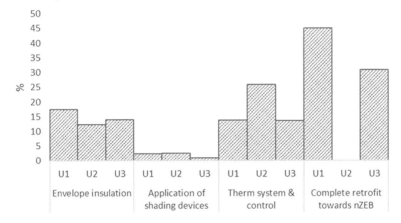

Fig. 8.3: RB in percentage compared to the new performance ($I_{RB,1}$), without dweller behavioral changes

Observing the $I_{RB,2}$ values, the gap between actual and estimated energy saving is higher for HAVC substitution by 35-55 per cent. In reverse, when energy efficiency improvements invest the envelope insulation or contemporarily the envelope and the thermal system, approximately 10 per cent of estimated energy saving is missing because of the rebound effect. The addition of solar shading devices as a single strategy is constantly inefficient, thus reducing the energy savings or openly do not arise as energy saving due to solar gains reduction in the cooling period are frustrated by higher heating consumption in the winter period (as fixed systems have been considered). In the depicted framework, higher dwellers' expectations after the renovation may produce negative $\otimes EP_{RB}$ (Fig. 8.4).

8.3 Impact of Occupancy in the Energy Behaviour of Educational Buildings

Occupation has a key role to play in the building performance about use,

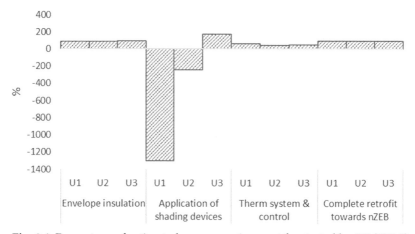

Fig. 8.4: Percentage of estimated energy savings not frustrated by RB (IRB,2)

energy, and comfort (IEA-EBC Annex 66). The occupancy unpredictability is part of the performance gap with the effect quantifiable around 20 per cent that can be moreover considerably heavier, depending on the building type (D'Oca *et al.*, 2018). The rate of occupancy and the user's influence on internal loads and ventilation frequency is frequently assumed as a standard value. Nevertheless for a comprehensive assessment of the building performance, detailed modelling approaches have been established and explored (Putra *et al.*, 2017; Chen *et al.*, 2018; Pan *et al.*, 2017).

In order to pursue the energy efficiency of the building and to control the balance between energy production and consumption in buildings, a consistent evaluation for an accurate load matching and for assessment of solar fraction (Voss *et al.*, 2010) turns out to be crucial.

8.3.1 The eLUX Lab as National Case Study

The eLUX lab building (Tagliabue *et al.*, 2015; De Angelis *et al.*, 2015) is the product of a multi-disciplinary research initiative at the smart campus of the University of Brescia, and it demonstrates smart technologies for energy retrofit in the educational 'learnscape'. The building was assumed as a pilot building of the S.C.U.O.LA (Smart Campus as Urban Open Lab) project in 2015, supported by a funding programme of Lombardy Regional and since 2016, it has become a strategic university project as 'eLUX – Energy Laboratory as University eXpo', headed by the Department of Information Engineering and partnered by the Department of Civil, Environmental, Architectural Engineering, and Mathematics, the Department of Mechanical and Industrial Engineering, the Department of Economics and Management, and the Department of Law.

Analysis of the building has been stated as the nationally funded PRIN project 'BHIMM – Modeling and management of information for the existing building stock' when the building was digitized, aimed at implementing energy management and promoting the introduction of renewable sources. During the PRIN project, a Building Information Model (BIM) was realized by collecting geometrical data by the use of a Terrestrial Laser Scanner (TLS) survey (Ciribini *et al.*, 2014).

In the first phase of the research, the digital model eased the conception of a virtual environment to manage, predict, and analyze the retrofit opportunities and the maintenance scenarios permitting data storage and organization of each building portion (e.g. envelope, thermal system, plumbing system, fire protection system, electrical system, and anti-intrusion system), as well as the zones-setting parameters adopted by the energy manager of the central university authority.

A Building Energy Model (BEM) was organized, starting from the digital model through the testing of interoperability paths, with the aim to envisage the results of a range of retrofitting measures for thermal and electrical energy saving. A first standard simulation showed the inadequacy to correctly define the thermal performance of the building, which had huge occupancy variability and multiple user-dependent factors (Tagliabue *et al.*, 2016). Consequently, the energy impact of the occupancy was explored (Zani *et al.*, 2016), turning the research towards cognitive building paradigm.

8.3.1.1 User-centered Design

The topic of user inclusion in the digital process promotes an approach to users' feedback as computable data, finally assuming the user as a human sensor to assess in a subjective way the indoor space conditions. The Distributed Sensing into eLUX lab allows detection of the indoor conditions through a monitoring and control system. However, the data provided by the sensors (i.e. people presence, temperature, humidity, illuminance, CO_2 concentration, VOC emissions, electricity consumption of ventilation systems fans) deliver objective data, describing environmental and operative conditions. On the other hand, the perception of the users and the option of enhancing the spaces' condition for users' support and well-being led to the requirement of customization of indoor conditions according to variable needs.

The eLUX lab is now an internationally known case study and an inter-departmental experimental field with multidisciplinary research, triggered by the vision of the working group and focusing on the concepts of sustainable energy, smart city, innovative technologies for predictive building management, distributed sensing, and thus, IoT security and privacy. The focus on the concept of connected and responsive building through smart applications for users' interaction defining the cognitive building concept has been developed (University of Brescia eLux website).

8.3.1.2 Cognitive Computing Approach

The cognitive building, based on cognitive computing (Chen *et al.*, 2018; Gupta *et al.*, 2018; Wang *et al.*, 2020), runs into a framework that joins a complete system architecture for building management and data processing. The data collection allows enhancement of the building efficiency through automation (Rinaldi *et al.*, 2016), allowing management of optimized rental processes and supporting financial compliance.

Furthermore, the better use of indoor spaces and customized control of the comfort conditions augments the user experience (UX) into the building.

The building becomes responsive and adapts itself to the user requirements, with a safe, personalized, and real-time service. Hereafter, the setup of the business models can be theoretically disruptive and crucial new-income forms could be branded for current resources.

The analysis of data collected by the users is used to set modulation, scheduled or predictive maintenance, on-demand services, which complete and integrate the process optimization (Pasini *et al.*, 2016).

Computable or cognitive computing data may include energy, water, lighting, indoor environmental condition, occupancy data, people flows, maintenance cycles, Enterprise Resource Planning (ERP) systems, rental contracts and financial information, IT devices, communication infrastructure, security and emergency systems, Building Management System (BMS) (Ploennigs *et al.*, 2017; Targowski, 2020).

The logic of cognitive management (Foteinos *et al.*, 2013; Groumpos and Mpelogianni, 2016; Wang, 2020) is reasonably reinforced by the establishment of a combined system for data management and gathering and which is able to track an information structure for process-control objectives.

The information model is thus the digital basis of this process as it is able to support, organize, and visualize the information by communicating (Shen *et al.*, 2012; Shen *et al.*, 2013) with simulation environments that can be used for specific detailed analysis, such as energy analysis, structural analysis, crowd simulation through gamification (Devis, 2016) or augmented reality through immersive environments (Mastrolembo Ventura *et al.*, 2016; Rossini *et al.*, 2016).

8.3.1.3 eLux Lab Spaces and Users

The eLUX lab building has three floors (i.e. underground, ground and first floor) where lecture rooms and computer labs are situated. A distribution atrium and an auditorium complete the asset. The building zones considered during the experimentations are labelled as given in Table 8.2. Numerous sensors are mounted in different zones, for different purposes, comprising energy monitoring, building automation, users' security, and many more. The IoT approach traced during the design and the preparation of the sensors network follows an easy combination of sensors, fitted to the suitable application domain. The complete building is equipped with more than one hundred measure points, using wireless communication equipment to limit the installation costs in the current situation.

8.3.2 Probabilistic Occupancy Profiles Test

The use of deterministic or probabilistic occupancy profiles considerably changes the result of the energy-consumption calculation. Accordingly, the possibility of identifying the occupancy outline through instruments in order to collect actual data on the operational use of the inner spaces is the decisive answer to adjust the predictive models to the influence of occupants on buildings' energy efficiency (Rinaldi *et al.*, 2018).

An example of this method can be found in the eLUX lab, the cognitive building in the smart campus at the University of Brescia. As first, it is worthy to note that the educational building has a demand in use and the impact of the users, in this case, the scholar is decisive (Tagliabue *et al.*, 2016). The eLUX lab has been adopted as a showcase through different national projects and it is the first Italian cognitive building applying a BIM-based methodology for energy renovation and developing an IoT network for internal checking.

In Fig. 8.5, thermal energy demand outlines for the eLUX lab are presented. The energy performance is mainly calculated, by using a standard profile (i.e. people density, internal gains, ventilation rate) for educational buildings in the national regulation context. The occupancy configurations have been determined by randomly varying the users/students attending value to the programmed lectures by a stochastic modeled progression. After that, the envelopment of the generated data outlines was plotted, picking up three stochastic profiles simulated on an hourly basis to wrap the possible energy loads as defined by probabilistic evaluation. In Fig. 8.5, the brown and orange lines describe the heating and cooling demands with the standard user's outline for the educational building. It manifests how these profiles of demands are always higher in the heating case, and lower in the cooling case, when compared to the wrapping shapes calculated through probabilistic occupancy outlines.

This means that the standard simulation underrates the energy demand in summer and overrates the energy demand in winter in the case of the university building, where the people flow has a major and unpredictable weight.

Table 8.2: Definition of Spaces and Occupants in eLUX Lab

			Floor					
		-1		0			1	
Name	*Unit*	*MLAB1*	*MLAB2*	*MTA*	*MTB*	*Atrium*	*M1*	
Area	[m^2]	152	208	178	177	181	337	
Volume	[m^3]	455	624	535	532	542	1012	
Occupants Standard	[n.]	76	104	89	89	90	262	
Actual	[n.]	56	82	168	168	56	169	

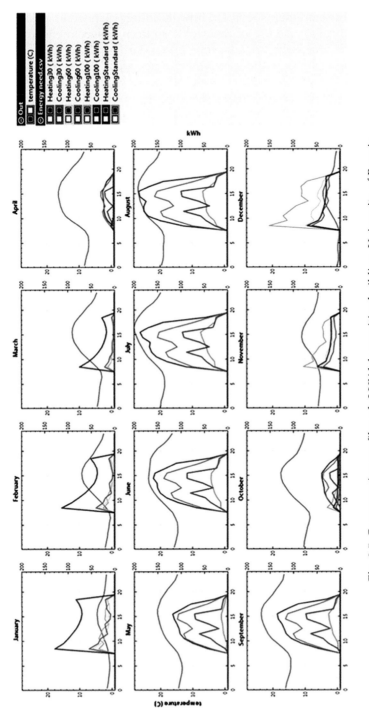

Fig. 8.5: Consumption profiles of eLUX lab cognitive building, University of Brescia

Many measurements revealed how buildings planned as zero-energy buildings through the use of renewable sources, failed to encounter the expectations precisely for adopting a standard occupancy simulation in the design phase, without assuming a specific user to determine, as an alternative, an unexpected usage of inside spaces. The appraisal of occupancy streams and the chance to accomplish a people tracking (Ciribini *et al.*, 2016) by an IoT network make a big difference in comparison with the standard occupancy that only normalizes the people density factor.

8.3.3 BIM-based Occupancy Flow Simulation

Occupancy flows can be calculated by predictive models (Yoo *et al.*, 2020; Di Giuda and Frate, 2020), which are able to delineate agents that simulate how users travel within the buildings' functional spaces for discovering where critical concerns may occur from distribution and movement point of view. This could be dangerous for evacuation events (Sun and Turkan, 2020) in buildings with high people density, wherever it is imperative to validate the linking between spaces and the possible accumulation of users to centralize services that could impact at organizational level (Sagun *et al.*, 2011). An additional aspect could be the modulation of the running of thermal service and comfort conditions (e.g. hygro-thermal, visual, indoor air quality, acoustic) that vary when different boundary conditions take turns.

A BIM-based (Building Information Model) analysis on the virtual spaces of the building permits to carry on a people flows reproduction through algorithms that use the asset geometry in order to compute the adjacencies of the spaces and the building features and, in conclusion, to hint the available alternatives of different paths for the agents that simulate the occurrence of the users moving within the spaces. In this way it is plausible to predict how people will act when dealing with route options and/or obstacles in a specific situation (Galland *et al.*, 2014).

Starting from the 3D BIM-based models, the method runs the users' interactions simulated by agents. Agents have size and acceleration, behaviors that can be defined, such as eluding ascending stairs (Banerjee and Kraemer, 2010). Agents can dynamically evaluate their location and make decisions about directions to follow with social forces governing how they answer to their surroundings, to other people and intentions. Agents try not to clash with other people and obstacles and they can make self-directed conclusions on the basis of an uninterrupted space (Szymanezyk *et al.*, 2011).

A first experiment on the simulation of occupancy flows was carried out on eLUX lab, the pilot building of the smart campus at the University of Brescia. The available as-built documents were reviewed and translated into a BIM model in Autodesk Revit-authoring tool, using a model definition level established according to the use of this specific case study. This meant the translation of Post-Occupancy Evaluation (POE) data into a dynamic replication which is able to support the analysis of existing occupancy situations. Spaces, circulation paths, external and internal walls were recreated to simulate end-user flows. The model was imported into the virtual environment of Mass Motion software (Morrow, 2011). The results shown in Fig. 8.6 and Fig. 8.7 are the evaluation of users flow and pathways of the students at the eLUX lab building to check the distribution density map of occupancy streams.

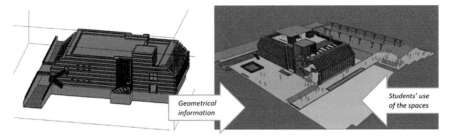

Fig. 8.6: BIM model and simulated people flow at eLUX lab

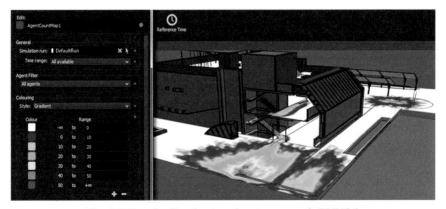

Fig. 8.7: People distribution density map of eLUX lab

The analyses were conducted on the eLUX lab in order to envisage the activities of virtual artificial intelligence agents and to simulate realistic qualitative images to assess and match primarily optional potential organizations and the variable educational methods.

A second level of analysis was performed in Unity 3D game engine to exemplify and simulate the usage scenario of the existing building. The definition of the use scenarios epitomizes the first step of a proof design phase based on the analysis of the end-users' behaviors. The replicated scenarios offer structured information on diverse alternative educational methods allocated in the building and representing a description of how the system can perform, giving the specific boundary conditions and defined at the proper level of definition which is required to achieve a consistent simulation result (Mastrolembo Ventura *et al.*, 2016).

The BIM model of the building was created and imported in Unity 3D and through Business Project Model and Notation (BPMN), the specifications of the use process were organized and executed in the virtual environment by the game engine (Fig. 8.8). The lecture rooms of the building are used for weekly educational activities and for the graduation days where the people flow grows and follows the academic ceremonies (Fig. 8.9).

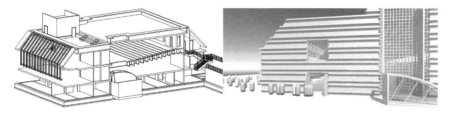

Fig. 8.8: BIM model of eLUX lab and simulation in the virtual environment Unity 3D

Fig. 8.9: Simulation of lecture days (*left*) and graduation days (*right*) in eLUX lab (a)

During the graduation days, the ventilation, comfort and Indoor Air Quality (IAQ) conditions are aggravated by the number of people rolling in the building spaces and fluctuating during the day according to the moment of discussion and during the academic proclamation. In the best scenario, the HVAC systems of the lecture rooms are designed on the basis of the number of seats; nevertheless, during the graduation days, every student has his/her parents and enthusiastic friends with a multiplier around 4 or more compared to the number of people attending the lectures during the regular week. (Fig. 8.9; Fig. 8.10).

In the distribution spaces during the weekly activities, the students move among the spaces to attend the lectures and to use the services located on the ground floor (females) and first floor (males).

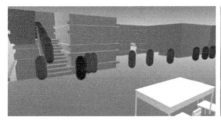

Fig. 8.10: Simulation of lecture days (*left*) and graduation days (*right*) in eLUX lab (b)

During the graduation days, the crowd of people waits in the atrium of the building to join the presentations in the lecture rooms and the charge of people (and consequently the internal gains and ventilation needs) in the distribution spaces and in the lecture rooms are intermittent at every 20-30 minutes according to the discussion schedule.

A correct approximation of the variables influencing the energy behavior of the building is a key factor in an effective methodology towards nZEB, both for new construction and energy upgrading. It is important to detect confidence verges and appropriate digital tools that can enable a level of accuracy in the predictive model verifications for delivering guaranteed results about the interaction users/building in terms of detailed energy use (internal loads, ventilation, set-point temperatures, use of shading devices, etc.) and space use. Scientific literature on this subject and the ongoing tests allow the definition of a Pre-Occupancy Evaluation (POE) scenario (Shen *et al.*, 2012; Schaumann *et al.*, 2019) that can also be carried out on the basis of the data gathered (Mastrolembo *et al.*, 2018) and kept updated during the lifecycle of the construction.

8.4   Cognitive Building

Usually the HVAC system may be attributed to increased consumptions where large public buildings are concerned and when the setup is not optimized. A traditional static control system which assumes an average occupancy percentage for every space is not appropriate when a cognitive building (Ploennig *et al.*, 2017) approach is pursued. The Energy Management System (EMS) can adjust the HVAC system embracing occupancy percentages based on predictive models which are able to modulate the quantity according to the users' routines and IAQ conditions supported by real-time data. The IoT network and the cognitive methodology have been tested in the eLUX lab and data delivered by IAQ sensors (i.e. temperature, relative humidity, CO_2, VOC) refine the results of the occupancy models of the spaces (Rinaldi *et al.*, 2018).

The optimized organization of the urban assets in a smart city approach involves the integration of the communication infrastructure

(Rinaldi *et al.*, 2017), data storage, and analysis systems (Bianchini *et al.*, 2018). The cognitive notion requires a building able to acquire the user preferences and lifestyle and that can cooperate with the sensors and building systems through actuators. A cognitive building minimizes its energy needs meanwhile preserving the users' comfort in a combined interaction between humans and asset. IAQ is among the main issues in several building uses and significantly disturbs the occupants' well-being, comfort, and efficiency, and also has a tough impact on the whole energy consumption (both thermal and electrical).

Actually, IAQ preservation is seen in educational buildings because of several aspects: high people concentration, lack of efficient air circulation, or absence of mechanical ventilation systems, which determine uncontrolled indoor conditions with unfitting IAQ levels with consequently, the incidence of Sick Building Syndrome (SBS) symptoms among sensible users (Takaoka *et al.*, 2016) and degradation of the learning performance (Mendell *et al.*, 2016).

Typically, carbon dioxide CO_2 is assumed as a meter of IAQ (Szczurek *et al.*, 2015) due to the easy detectability through sensors (Torresani *et al.*, 2013).

8.4.1 IoT Network to Detect Uncomfortable Conditions

In the eLUX lab, the testing is piloted to examine the possibility to mix IAQ data coming from IoT sensors to upturn the details on occupancy evaluation in the building spaces (Casado-Vara *et al.*, 2018), with the aim to finally adjust the HVAC system.

Implementation of the IoT architecture (Akkaya *et al.*, 2015; Pan *et al.*, 2015) for the sensors' interconnection to the building management system has been explored by eLUX researchers (Bellagente *et al.*, 2015) and the method has been approved to progress in the precision of the occupancy assessment of probabilistic models, as shown in the previous section.

8.4.1.1 Ventilation Rate for Indoor Spaces

Ventilation of the indoor space is based on minimum ventilation per use, people density, dimension and volume of the space as defined in the national and international standards (ASHRAE, 62-89; UNI, 10339).

The numbers n of air changes per hour $[h^{-1}]$ are therefore calculated according to the following equation (4):

$$n = \frac{(v_{min} \cdot i_s \cdot A)}{V} \tag{4}$$

where: n is the number of air changes; v_{min} is the required external air $[m^3/h$ per person]; i_s is the people density $[person/m^2]$; A is the area of the indoor space $[m^2]$, and V is the volume $[m^3]$.

In educational spaces of every level, the national standard attributes the value 0.5 to i_s and v_{min} is 21.6 m³/h per person.

8.4.1.2 IAQ Calculation and Definition

The IAQ calculation recommends a minimum ventilation rate to guarantee the right IAQ in an indoor space of a certain size and where a specific activity takes place.

The human activities produce indoor CO_2 through users' breath and the ventilation rate has to eliminate the CO_2 concentration to promote an acceptable level (stated at 1000 ppm). The value c is the CO_2 concentration [m³/m³] during time t [h] in a thermal zone calculated according to the equation (5) that includes n and V defined in equation (4):

$$c = \left(\frac{q}{nV}\right) \cdot \left[1 - \left(\frac{1}{e^{nt}}\right)\right] + (c_0 - c_1) \cdot \left(\frac{1}{e^{nt}}\right) + c_i \tag{5}$$

where: q is the quantity of the CO_2 produced by people [m³/h]; c_o is the CO_2 concentration at the starting point $t = 0$ of the calculation and set at 350 ppm; c_i is the CO_2 concentration in the inlet pipe, set at 0 m³/m³. In the indoor space, the CO_2 generation is quantified according to equation (6):

$$q = q_p \cdot n_0 \tag{6}$$

where: q is the CO_2 exhaled by every person in the room activity (0.05 m³/h/person) [m³/h]; n_o is the people number [-]. The effect of CO_2 concentration on IAQ condition is defined in Table 8.3 (Pietrucha, 2015).

The upper limit of fresh air is 1000 ppm, however 600 ppm is the satisfactory level in learning space to preserve optimal and healthy indoor conditions when social protection is a priority.

8.5 Methodology for a Cognitive Strategy

8.5.1 Inverse Modeling Through Data Gathering

The validation of energy models is conceivable merely by actual behavior monitoring, otherwise the unavoidable performance gap between predictive model and actual data is not overwhelmed (Plageras *et al.*, 2018). In the eLUX lab, people counters are mounted together with CO_2 and VOC concentration sensor. The CO_2 concentration sensors are essential to delineate the occupancy rate and are used to digitally control through direct digital control (DDC) the mechanical ventilation system to modify the airflow of the air handling units (AHUs).

Additionally, inverse calculation of the occupancy percentage can be completed based on CO_2 concentration and additional inverse modeling approaches are accessible, such as multiple linear regression models for

Table 8.3: Indoor Air Quality Related to CO_2 Concentration

	CO$_2$ Concentration [ppm]					
	350–400	*<600*	*<1000*	*<1500*	*<2000*	*<10000*
IAQ	Fresh air	Nearly fresh air	Upper limit	Stuffy air	Weak people can faint and cough	Increase in breath rate, respiratory problems, headaches, nausea
Condition	Healthy	Acceptable	Limit	Not acceptable	Bad	Very bad

the characterization of drivers factors (Tronchin *et al.*, 2016) or as the aim is to gradually adjust the energy models (Rodríguez *et al.*, 2013).

8.5.2 Monitoring of the Indoor Conditions

In the eLUX lab, the fixing of CO_2 concentration, temperature, and relative humidity sensors, allows to weigh the indoor comfort condition and IAQ considering that the latter dramatically deteriorates as the temperature and relative humidity rise. The parameters and the values set out in the standards (UNI/TS, 11300; UNI, 10349-1/2/3) for category E.7 (D.P.R., 412/93) which includes university lecture rooms are précised in the Table 8.4.

Table 8.4: Indoor Comfort and IAQ Parameters

Parameter	Unit	Winter	Summer
Dry bulb external temp.[T_{bse}]	°C	-7	32
External relative humidity [UR_e]	%	60	48
Dry bulb indoor temp. [T_{bsa}]	°C	≤ 20	≥ 26
Indoor relative humidity [UR_a]	%	$35<UR_a<45$	$50<UR_a<60$
Metabolic activity [Mr]	W/m^2	≥ 70	≤ 116
Thermal resistance of clothing [Icl]	m$^{2\circ}$C/W	≥ 0.14	≤ 0.09
Air velocity [v]	m/s	$0.05<v<0.15$	$0.05<v<0.15$

8.5.3 IoT for Cognitive Implementation

Commonly IAQ sensors are coupled with control HVAC system with no data transfer for different applications. The IoT paradigm overcomes this problem, by organization of an appropriate ICT architecture accomplished to virtualize the physical devices through a specific data model. The ICT architecture distributes the different layers from device layer at the bottom, to the databases in the data layer, device data, BIM or AMS (Asset Management System) to the cognitive layer for implementing the processes of learning and of sending commands to actuators, and finally to the visualization layer (Fig. 8.11).

The device layer comprises sensors and actuators situated in the field and characterized by data models which are able to communicate through proper protocols.

The system can support numerous communication protocols, including MQTT (MQ Telemetry Transport or Message Queue Telemetry Transport) and web service, REST. The data collected are stored in the databases of the data layer which encompasses the run-time data from sensors (device data DB) as well as data about the building (BIM DB),

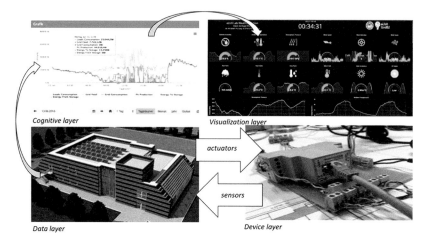

Fig. 8.11: Architecture layers to control parameters for the cognitive
approach implementation

and data about the devices connection and their features (Asset mgt DB).
The cognitive layer can use the information of the device data DB to
extract additional information (i.e. the number of people in a space from
CO_2 concentration). The IoT method has the main benefit to allow the
opportunity to combine sensors from a number of domains, as well as the
virtual one and in the organization of the eLUX lab the estimated data are
uploaded in the Device Data, as a virtual sensor by the cognitive layer.

8.5.4 Occupancy Scenarios

The number of students in the eLUX lab has been confirmed through an
inspection on the actual availability of seats in the lecture rooms, finding
that the standard density is not always respected (Table 8.2). The lecture
schedule is actually highly intense during the weekdays.

The results about the standard energy analysis and the probabilistic
calculation were previously exposed. In this second step, the analysis
focuses on the indoor conditions and the endorsement of the probabilistic
predictive model on IAQ and indoor comfort conditions through
Distributed Sensing (DS) method.

8.6 Results

The indoor conditions were replicated according to the occupancy scenario
as summarized in Table 8.5. The air changes provided by the installed Air
Handling Units (AHUs) are reported (1) together with the standard air
changes based on the regulation (2) and with the air changes required to
achieve the CO_2 threshold of 1000 ppm (3).

Table 8.5: Air Change Gaps among Standard Calculation/Actual Occupancy Situation Needed for IAQ

		Floor					
		-1		0		1	
Name	Unit	MLAB1	MLAB2	MTA	MTB	Atrium	M1
(1) AHU	[m³/h]	5000	2000	2000	2000	5500	5500
(2) Standard	[m³/h]	1639	2246	1925	1917	1952	3645
(3) IAQ	[m³/h]	1427	2786	4281	4281	1903	6677
$Gap_{standard}$	[m³/h]	3361	-246	75	83	3548	1855
Gap_{iaq}	[m³/h]	3573	-786	-2281	-2281	3597	-1177

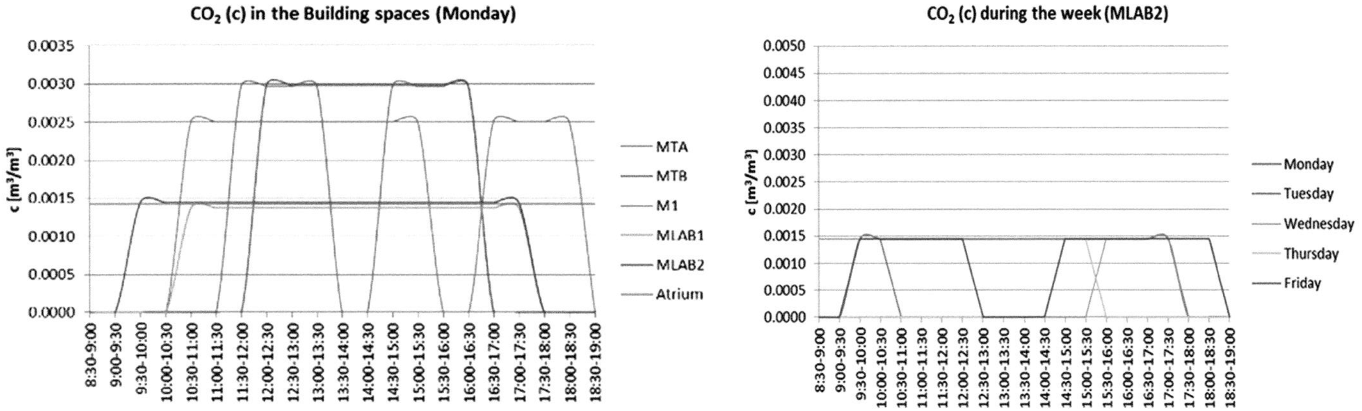

Fig. 8.12: Predicted CO_2 concentration in eLUX lab learning spaces

The gap among the AHUs' capacity and the air changes in the two scenarios are shown to emphasize the granularity and weight of the concern in the different eLUX lab spaces. The negative values in the Gap_{iaq} and $Gap_{standard}$ give advice that the actual AHUs are supporting the real number of people attending the lectures when the occupancy rate goes above the standard value.

The predictive model about CO_2 concentration in the lecture rooms shows high discomfort level due to exceeded IAQ limits with effects of tiredness for the students and olfactory discomfort. In the rooms where outdated AHUs are mounted (only in the ground floor AHUs were substituted in 2010) the level of CO_2 can touch 3000 ppm during the weekdays. Tolerable levels can be calculated for the atrium and the computer lab on the underground floor (Fig. 8.12).

The IoT data validated the predictive model and a test room was designated for the proof, i.e. the MLAB2 situated in ground floor (Table 8.2). The IAQ limit and the temperature and relative humidity were checked in the room by Siegenia Sensoair for CO_2 and VOC detection and Everspring ST814 temperature and relative humidity sensor. The sensors supported Z-wave communication protocol and they are connected with the monitoring system through a dedicated gateway.

A three-month test period was analyzed and the collected values matched with data reported in Table 8.4 that represents the compliant conditions for educational spaces. The thresholds defined by the standards are illustrated in diagrams to show how many value among the recorded samples drop in the comfort range. The diagrams show the Cumulative Distribution Function (CDF) for each month in the monitoring campaign. Finally a test day is used to exemplify the thermo-hygrometric comfort and IAQ conditions coupled with estimation of the number of occupants elaborated by the IAQ sensor. This result confirms likewise the probabilistic predictive model of occupancy embraced for the building and described to assess the fluctuation of energy need related to occupancy variation.

In Fig. 8.13, the temperature in the MLAB2 is plotted and the 27 per cent of data is lower than the winter comfort mainly due to the public holiday period included in the dataset. Almost 70 per cent of the recorded data surpasses the standard limit of 20°C because the room is a computer lab and for this reason, the internal gains for PC, artificial lighting, and students participating at the lectures and working for their exercises is extremely high. The green area superposed on the diagram of relative humidity delimits the comfort range but almost 80 per cent of the values are out of range, frequently in the dry air section.

In Fig. 8.14 the Cumulative Distribution Function of CO_2 shows that when the room is used by the students, the indoor contamination surpasses the IAQ border until 1400 ppm with the effect of perception of stinking, and not fresh air. In a typical week day at eLUX lab more than

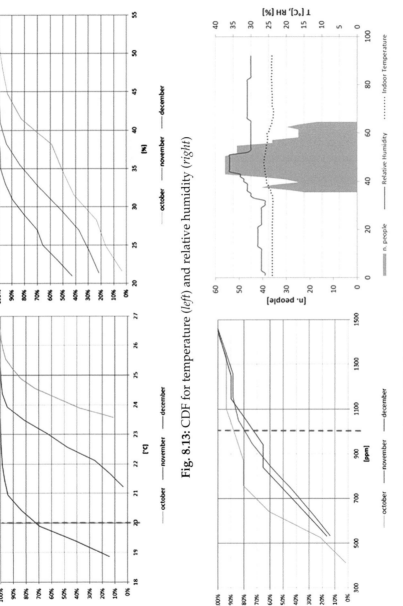

Fig. 8.13: CDF for temperature (*left*) and relative humidity (*right*)

Fig. 8.14: CDF for the CO_2 (*left*) and indoor conditions for a test day (*right*).

50 per cent of the values of CO_2 concentration overcome 1000 ppm with a maximum value of 1500 ppm, confirming the air change gap expected by the predictive model (Fig. 8.12). During the typical weekly day, the students' number increases in the morning and the relative humidity is lower than 35 per cent in the 90 per cent of the records with the temperature constantly over 20°C even during the night. The estimated number of students are consistent with CO_2 concentration and endorse the average percentage of attendance to the lectures of 60 per cent, validating the probabilistic occupancy model (Tagliabue *et al.*, 2016) proposed in Fig. 8.5.

Therefore, the IoT network of sensors, validated and allowed to adjust the predictive models for evaluating the building behavior and adaptive strategies based on BMS to adapt HVAC and AHUs flows, can be based on the real-time information provided by the network.

8.7 Conclusion

The opportunity to plot the comfort data into the BIM model and to link the IoT network to a digital database correlated to the building improve the daily management of the internal spaces, allowing regulation of the settings according to flexible people flows during the day with a twofold outcome – dropping the energy waste, and expanding the comfort condition, and the UX (Fig. 8.15).

The cognitive paradigm could incorporate the user as a sensor and therefore promote a data crossing between sensors and users' perception

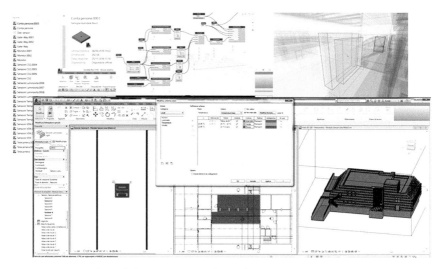

Fig. 8.15: BIM-based data mapping of comfort parameters and IAQ to detect management issues

that could be affected by the position in the space, proximity to vents, equipment or to windows when operable. The testing showed at eLUX, indicated as the IoT architecture, can exploit the use of sensors data for different application domains and how real-time evaluation of the indoor condition and occupancy percentage can allow automatic learning procedure in artificial intelligence (AI) and machine learning (ML), extending the opportunity to establish a synergy among the systems of the building (e.g. shading devices, HVAC, windows opening, but also electrical generation and storage systems, etc.) and of the smart city where the flow of information needs to be managed to regulate and optimize the complex urban structure.

Furthermore, the occupancy fluctuation and the impact in the energy outcome require a detailed modeling and evaluation of the people stream that could be embraced to predict future renovation scenarios for the existing buildings or POE strategies for new constructions through BIM-based technologies which are able to encourage inclusive design processes in collaborative environments, such as AR (augmented reality) or VR (virtual reality). The crowd simulation could be also be protracted from fire safety to increase the UX into the building, supporting a gradual improvement of positive results on public health and well-being in spaces dialoguing with humans.

The power of forecast and anticipation and the opportunity to corroborate the models demonstrating actual processes through linked objects will develop a smart built environment that essentially could sustain social, environmental, and economic equity in future cities.

References

Akkaya, K., Guvenc, I., Aygun, R., Pala, N. and Kadri, A. (2015). IoT-based occupancy monitoring techniques for energy-efficient smart buildings. *In*: 2015 IEEE Wireless Communications and Networking Conference Workshops (WCNCW), pp. 58-63.

ASHRAE, 62-89. (1989). *Ventilation for Acceptable Indoor Air Quality*. American Society of Heating, Refrigerating and Air-conditioning Engineers, Inc. Atlanta, US.

Banerjee, B. and Kraemer, L. (2010). Evaluation and comparison of multi-agent based crowd simulation systems. *In*: International Workshop on Agents for Games and Simulations, pp. 53-66, Springer, Berlin, Heidelberg.

Bellagente, P., Ferrari, P., Flammini, A. and Rinaldi, S. (2015). Adopting IoT framework for Energy Management of Smart Building: A real test-case. *In*: Proc. of IEEE RTSI, Turin, Italy, Sept. 16-18, pp. 138-143.

Bianchini, D., De Antonellis, V., Melchiori, M., Bellagente, P. and Rinaldi, S. (2018). Data management challenges for smart living. Lecture Notes of the Institute for

Computer Sciences, Social-Informatics and Telecommunications Engineering, LNICST, pp. 131-137.

Bosia, D. (2013). Traditional built environment conservation: Social and technological aspects. *Smart and Sustainable Built Environments*, 457.

Butera, F.M. (2013). Zero-energy buildings: The challenges. *Advances in Building Energy Research*, 7(1): 51-65.

Calì, D., Osterhage, T., Streblow, R. and Müller, D. (2016). Energy performance gap in refurbished German dwellings: Lesson learned from a field test. *Energy and Buildings*, 127: 1146-1158.

Casado-Vara, R., Vale, Z., Prieto, J. and Corchado, J.M. (2018). Fault-tolerant temperature control algorithm for IoT networks in smart buildings. *Energies*, 11(12): 3430.

Chen, M., Herrera, F. and Hwang, K. (2018). Cognitive computing: Architecture, technologies and intelligent applications. *IEEE Access*, 6: 19774-19783.

Chen, Y., Hong, T. and Luo, X. (2018). An agent-based stochastic occupancy simulator. *Building Simulation*, 11(1): 37-49. Tsinghua University Press.

Cheng, J.C., Chen, K. and Chen, W. (2020). State-of-the-art review on mixed reality applications in the AECO industry. *Journal of Construction Engineering and Management*, 146(2): 03119009.

Ciribini, AL.C., Pasini, D., Tagliabue, L.C., Manfren, M., Daniotti, B., Rinaldi, S. and De Angelis, E. (2016). Tracking users' behaviors through real-time information in BIMs: Workflow for interconnection in the Brescia Smart Campus Demonstrator. International High Performance Built Environment Conference – A Sustainable Built Environment Conference 2016 Series (SBE16). iHBE 2016, *Procedia Engineering*, 180(2017): 1484-1494.

Ciribini, A.L.C., Mastrolembo Ventura, S. and Paneroni, M. (2014). La metodologia BIM a sostegno di un approccio integrato al processo conservativo. *In*: Stefano della Torre (Ed.). Preventive and Planned Conservation, vol. 5, ICT per il miglioramento del processo conservativo; Proc. Intern. Symp, Monza-Mantova, 5-9 Maggio 2014, Nardini Editore.

Corrado, V., Ballarini, I., Paduos, E. and Primo, E. (2016). The rebound effect after the energy refurbishment of residential buildings towards high performances. International High Performance Buildings Conference, July 11-14, Purdue University. http://docs.lib.purdue.edu/ihpbc/180

Cuerda, E., Guerra-Santin, O., Sendra, J.J. and Neila, F.J. (2020). Understanding the performance gap in energy retrofitting: Measured input data for adjusting building simulation models. *Energy and Buildings*, 209: 109688.

Davis, D. (2016). Evaluating Buildings with Computation and Machine Learning. *In*: ACADIA 2016 (Ed.). Velikov, Sandra Manninger, Matias del Campo, Sean Ahlquist, and Geoffrey Thun, 116-23. Ann Arbour, Michigan.

De Angelis, E., Ciribini, A., Tagliabue, L.C. and Paneroni, M., (2015). The Brescia Smart Campus Demonstrator. Renovation toward a zero energy classroom building. *Procedia Engineering*, 118: 735-743.

De Wilde, P. (2014). The gap between predicted and measured energy performance of buildings: A framework for investigation. *Automation in Construction*, 41: 40-49.

Di Giuda, G.M. and Frate, M. (2020). Use of predictive analyses for BIM-based space quality optimization: A case study. Progetto Iscol. *In*: Buildings for Education, pp. 185-192. Springer, Cham.

Directive 2010/31/EU of The European Parliament and of The Council of 19 May 2010 on the energy performance of buildings (recast). *Official Journal of the European Union*, L 153/13, 18.6.2010.

Directive 2012/27/EU of The European Parliament and of The Council of 25 October 2012 on energy efficiency, amending Directives 2009/125/EC and 2010/30/EU and repealing Directives 2004/8/EC and 2006/32/EC (Text with EEA relevance). *Official Journal of the European Union*, L 315/1, 14.11.2012.

Directive (EU) 2018/844 of The European Parliament And of The Council of 30 May 2018 amending Directive 2010/31/EU on the energy performance of buildings and Directive 2012/27/EU on energy efficiency (Text with EEA relevance). *Official Journal of the European Union*, L 156/75, 19.6.2018.

D'Oca, S., Hong, T. and Langevin, J. (2018). The human dimensions of energy use in buildings: A review. *Renewable and Sustainable Energy Reviews*, 81: 731-742.

D.P.R. (26 Agosto 1993). n. 412. Regolamento recante norme per la progettazione, l'installazione, l'esercizio e la manutenzione degli impianti termici degli edifici ai fini del contenimento dei consumi di energia. In: Attuazione dell'art. 4, comma 4.

Figueiredo, A., Kämpf, J., Vicente, R., Oliveira, R. and Silva, T. (2018). Comparison between monitored and simulated data using evolutionary algorithms: Reducing the performance gap in dynamic building simulation. *Journal of Building Engineering*, 17: 96-106.

Foteinos, V., Kelaidonis, D., Poulios, G., Vlacheas, P., Stavroulaki, V. and Demestichas, P. (2013). Cognitive management for the internet of things: A framework for enabling autonomous applications. *IEEE Vehicular Technology Magazine*, 8(4): 90-99.

Galland, S., Gaud, N., Demange, J. and Koukam, A. (2014). Multilevel model of the 3D virtual environment for crowd simulation in buildings. *Procedia Computer Science*, 32: 822-827.

Galvin, R. (2014). Making the 'rebound effect' more useful for performance evaluation of thermal retrofits of existing homes: Defining the 'energy savings deficit' and the 'energy performance gap'. *Energy and Buildings*, 69: 515.

Groumpos, P.P. and Mpelogianni, V. (2016). An overview of fuzzy cognitive maps for energy efficiency in intelligent buildings. *In*: 2016 7th International Conference on Information, Intelligence, Systems & Applications (IISA), pp. 1-6, IEEE.

Gupta, S., Kar, A.K., Baabdullah, A. and Al-Khowaiter, W.A. (2018). Big Data with cognitive computing: A review for the future. *International Journal of Information Management*, 42: 78-89.

Haas, R. and Biermayr, P. (2000). The rebound effect for space heating: Empirical evidence from Austria. *Energy Policy*, 28(6-7): 403-410.

Hijazi, M.O. (2015). *Bridging the Gap: A Tool to Support BIM Data Transparency for Interoperability with Building Energy Performance Software*. University of Southern California.

IEA-EBC. Annex 66 – International Energy Agency/Energy in Building and Community Programme: Definition and Simulation of Occupant Behavior in Buildings. https://annex66.org/ (Retrieved 03/12/2020)

Imam, S., Coley, D.A. and Walker, I. (2017) The building performance gap: Are modelers literate? *Building Serv. Eng. Res. Technol.*, F 0(0): 1-25. doi: 10.1177/0143624416684641

Ioannidis, D., Vidaurre-Arbizu, M., Martin-Gomez, C., Krinidis, S., Moschos, I., Zuazua-Ros, A. and Likothanassis, S. (2017). Comparison of detailed occupancy profile generative methods to published standard diversity profiles. *Personal and Ubiquitous Computing*, 21(3): 521-535.

Johnston, D., Farmer, D., Brooke-Peat, M. and Miles-Shenton, D. (2016). Bridging the domestic building fabric performance gap. *Building Research and Information*, 44(2): 147-159.

Kampelis, N., Gobakis, K., Vagias, V., Kolokotsa, D., Standardi, L., Isidori, D. and Venezia, L. (2017). Evaluation of the performance gap in industrial, residential & tertiary near-zero energy buildings. *Energy and Buildings*, 148: 58-73.

Liang, J., Qiu, Y. and Hu, M. (2019). Mind the energy performance gap: Evidence from green commercial buildings. *Resources, Conservation and Recycling*, 141: 364-377.

Lu, Y., Zhang, N. and Chen, J. (2017). A behavior-based decision-making model for energy performance contracting in building retrofit. *Energy and Buildings*, 156: 315-326.

Mastrolembo Ventura, S., Hilfert, T., Archetti, M., Rizzi, M., Spezia, A., Tagliabue, L. and Ciribini, A. (2018). Evaluation of building use scenarios by crowd simulations and immersive virtual environments: A case study. *In*: 35th International Symposium on Automation and Robotics in Construction and Mining (ISARC 2018), pp. 1-9.

Mastrolembo Ventura, S., Ghelfi, D., Oliveri, E. and Ciribini, A.L.C. (2016). Proceedings from ISTeA 2016, Back to 4.0. Rethinking the Digital Construction Industry: Building Information Modeling and Gamification for Educational Facilities. Maggioli Editore. ISBN:978-88-916-1807-8

Mastrolembo Ventura, S., Simeone, D., Ghelfi, D, Oliveri, E. and Ciribini, A.L.C. (2016). Building Information Modeling and Gamification for educational facilities. ISTeA 2016, back to 4.0. Rethinking the Digital Construction Industry. Maggioli Editore, 2016, pp. 203-212, ISBN 978-88-916-1807-8

Mendell, M.J., Eliseeva, E.A., Davies, M.M. and Lobschied, A. (2016). Do classroom ventilation rates in California elementary schools influence standardized test scores? *Results from a Prospective Study*, Indoor Air, 26: 546-557.

Menezes, A.C., Cripps, A., Bouchlaghem, D. and Buswell, R. (2012). Predicted vs. actual energy performance of non-domestic buildings: Using post-occupancy evaluation data to reduce the performance gap. *Applied Energy*, 97: 355-364.

Morrow, E. (2011). Efficiently using micro-simulation to inform facility design – A case study in managing complexity. *In*: Pedestrian and Evacuation Dynamics, pp. 855-863. Springer, Boston, MA.

Oliver, S., Seyedzadeh, S. and Rahimian, F. (2019). Using real occupancy in retrofit decision-making: Reducing the performance gap in low utilisation higher education buildings. In: 36th CIB W78 2019 Conference: ICT in Design, Construction and Management in Engineering, Architecture, Engineering, Construction and Operations (AECO), 676, University of Northumbria, Newcastle-upon-Tyne.

Pan, J., Jain, R., Paul, S., Vu, T., Saifullah, A. and Sha, M. (2015). An Internet of

Things framework for smart energy in buildings: Designs, prototype and experiments. *IEEE Internet of Things Journal*, 2(6): 527-537.

Pan, S., Wang, X., Wei, Y., Zhang, X., Gal, C., Ren, G. and Xie, J. (2017). Cluster analysis for occupant-behavior based electricity load patterns in buildings: A case study in Shanghai residences. *Building Simulation*, 10(6): 889-898, Tsinghua University Press.

Pasini, D., Mastrolembo Ventura, S., Rinaldi, S., Bellagente, P., Flammini, A. and Ciribini, A.L.C. (2016). Exploiting Internet of Things and Building Information Modeling Framework for Management of Cognitive Buildings, IEEE Second International Smart Cities Conference (ISC2 2016), Improving the Citizens' Quality of Life, 12-15 September, Trento, Italy, Proceedings IEEE, pp. 478-483, ISBN: 978-1-5090-1845-1.

Perez-Lombard, L., Ortiz, J. and Pout, C. (2008) A review on buildings energy consumption information. *Energy and Buildings*, 40(3): 394-398.

Pietrucha, T. (2015). Measurement of carbon dioxide concentration for assessment of indoor air quality in the lecture hall. 13th Students' Science Conference, Polanica-Zdrój Poland, 17-20 September.

Plageras, A.P., Psannis, K.E., Stergiou, C., Wang, H. and Gupta, B.B. (2018). Efficient IoT-based sensor BIG Data collection-processing and analysis in smart buildings. *Future Generation Computer Systems*, 82: 349-357.

Ploennigs, J., Ba, A. and Barry, M. (2017). Materializing the promises of cognitive IoT: How cognitive buildings are shaping the way. *IEEE Internet of Things Journal*, 5(4): 2367-2374.

Putra, H.C., Andrews, C.J. and Senick, J.A. (2017). An agent-based model of building occupant behavior during load shedding. *Building Simulation*, 10(6) 845-859, Tsinghua University Press.

Re Cecconi, F., Manfren, M., Tagliabue, L.C., Marenzi, G., De Angelis, E., Ciribini, A.L.C. and Zani, A. (2017). *In*: Proceedings IBPSA International Building Performance Simulation Association Conference: Surrogate Models to Cope with Users' Behaviour in School Building Energy Performance Calculation, (Ed). Charles S. Barnaby and Michael Wetter (Ed.). ISBN: 978-1-7750520-0-5.

Rinaldi, S., Flammini, A., Tagliabue, L.C., Ciribini, A.L.C., Pasetti, M. and Zanoni, S. (2018). Metrological Issues in the Integration of Heterogeneous Iot Devices for Energy Efficiency in Cognitive Buildings. I2MTC – 2018 IEEE International Instrumentation & Measurement Technology Conference, May 14-16, 2018. Royal Sonesta Hotel, Houston, Texas, USA.

Rinaldi, S., Flammini, A., Tagliabue, L.C. and Ciribini, A.L.C. (2018) On the use of IoT Sensors for Indoor Conditions Assessment and Tuning of Occupancy Rates Models, IEEE International Workshop on Metrology for Industry 4.0 and IoT, 16-18 April. Brescia, Italy.

Rinaldi, S., Bonafini, F., Ferrari, P., Flammini, A. and Rizzi, M. (2017) Evaluating low-cost bridges for time sensitive software defined networking in smart cities. *In*: Proc. of IEEE ISPCS, USA. Aug. 27-Sep. 1, pp. 7-12.

Rinaldi, S., Bittenbinder, F., Che, L., Bellagente, P., Tagliabue, L.C. and Ciribini, A.L.C. (2016). Bi-directional Interactions between Users and Cognitive Buildings by means of Smartphone App. IEEE Second International Smart Cities Conference (ISC2 2016) Improving the Citizens Quality of Life, 12-15 September 2016. Trento, Italy, Proceedings IEEE, pp. 490-495, ISBN: 978-1-5090-1845-1.

Rodríguez, G.C., Andrés, A.C., Muñoz, F.D., López, J.M.C. and Zhang, Y., (2013). Uncertainties and sensitivity analysis in building energy simulation using macroparameters. *Energy and Buildings*, 67: 79-87.

Rossini, F.L., Fioravanti, A., Novembri, G. and Insola, C., (2016). HOLOBUILD: Process optimization by the introduction of Mixed Reality in construction site. ISTeA 2016, BACK TO 4.0, Rethinking the Digital Construction Industry. Maggioli Editore, 2016, pp. 279-288, ISBN 978-88-916-1807-8.

Sagun, A., Bouchlaghem, D. and Anumba, C.J. (2011). Computer simulations vs. building guidance to enhance evacuation performance of buildings during emergency events. *Simulation Modelling Practice and Theory*, 19(3): 1007-1019.

Santos, R.S., Matias, J.C., Abreu, A. and Reis, F. (2018). Evolutionary algorithms on reducing energy consumption in buildings: An approach to provide smart and efficiency choices, considering the rebound effect. *Computers & Industrial Engineering*, 126: 729-755.

Seebauer, S. (2018). The psychology of rebound effects: Explaining energy efficiency rebound behaviours with electric vehicles and building insulation in Austria. *Energy Research and Social Science*, 46: 311-320.

Shen, W., Xiaoling Z., Shen, G.Q. and Fernando, T. (2013). The user pre-occupancy evaluation method in designer-client communication in early design stage: A case study. *Automation in Construction*, 32: 112-124.

Shen, W., Shen, Q. and Sun, Q. (2012). Building information modeling-based user activity simulation and evaluation method for improving designer-user communications. *Automation in Construction*, 21: 148-160.

Shen, W., Shen, G.Q. and Xiaoling, Z. (2012). A user pre-occupancy evaluation method for facilitating the designer-client communication. *Facilities*, 30.7/8: 302-323.

Shi, X., Si, B., Zhao, J., Tian, Z., Wang, C., Jin X. and Zhou, X. (2019). Magnitude, causes, and solutions of the performance gap of buildings: A review. *Sustainability*, 11(3): 937.

Schaumann, D., Pilosof, N.P., Sopher, H., Yahav, J. and Kalay, Y.E. (2019). Simulating multi-agent narratives for pre-occupancy evaluation of architectural designs. *Automation in Construction*, 106: 102896.

Sun, Q. and Turkan, Y. (2020). A BIM-based simulation framework for fire safety management and investigation of the critical factors affecting human evacuation performance. *Advanced Engineering Informatics*, 44: 101093.

Szczurek, A., Maciejewska, M. and Pietrucha, T. (2015). CO_2 and volatile organic compounds as indicators of IAQ. 6th AIVC Conference, 5th Tight Vent Conference, 3rd Venticool Conference, Madrid, Spain. 23-24 September.

Szymanezyk, O., Dickinson, P. and Duckett, T. (2011). Towards agent-based crowd simulation in airports using games technology. *In*: KES International Symposium on Agent and Multi-Agent Systems: Technologies and Applications, pp. 524-533. Springer, Berlin, Heidelberg.

Tagliabue, L.C., Manfren, M., Ciribini, A.L.C. and De Angelis, E. (2016). Probabilistic behavioural modeling in building performance simulation – The Brescia eLUX lab. *Energy and Buildings*, 128: 119-131.

Tagliabue, L., Manfren, M. and De Angelis, E. (2015). Energy efficiency assessment-based on realistic occupancy patterns obtained through stochastic simulation. pp. 469-478. *In*: M.R. Thomsen, M. Tamke, C. Gengnagel, B. Faircloth, F. Scheurer (Eds.). Modeling Behaviour, Springer International Publishing, 2015.

Takaoka, M., Suzuki, K. and Norback, D. (2016). Sick building syndrome among junior high school students in Japan in relation to the home and school environment. *Global Journal of Health Science*, 8(2): 165-177.

Targowski, A. (2020). Cognitive informatics and wisdom development: Interdisciplinary approaches. *Additive Manufacturing*, 2021: 949.

Thomas, S. and Rosenow, J. (2020). Drivers of increasing energy consumption in Europe and policy implications. *Energy Policy*, 137: 111108.

Torresani, W., Battisti, N., Maglione, A., Brunelli, D. and Macii, D. (2013). A multi-sensor wireless solution for indoor thermal comfort monitoring. *In*: Proc. of IEEE EESMS, pp. 25-30. Trento, Italy, 11-12 Sep.

Tronchin, L., Manfren, M. and Tagliabue, L.C. (2016). Optimization of building energy performance by means of multi-scale analysis – Lessons learned from case studies. *Sustainable Cities and Society*, 27: 296-306.

UNI 10339 (1995). Impianti aeraulici a fini di benessere – Generalità, classificazione e requisiti – Regole per la richiesta dell'offerta, l'offerta, l'ordine e la fornitura.

UNI 10349-1 (2016). Riscaldamento e raffrescamento degli edifici – Dati climatic, Parte 1: Medie mensili per la valutazione della prestazione termo-energetica dell'edificio e metodi per ripartire l'irradianza solare nella frazione diretta e diffusa e per calcolare l'irradianza solare su di una superficie inclinata.

UNI 10349-3 (2016). Riscaldamento e raffrescamento degli edifici – Dati climatici - Parte 3: Differenze di temperatura cumulate (gradi giorno) ed altri indici sintetici. Università degli Studi di Brescia, Progetto eLUX – Energy Laboratory as Univerity eXpo. Available online: http://elux.unibs.it/

UNI/TS 11300-1 (2014). Prestazioni energetiche degli edifici – Parte 1: Determinazione del fabbisogno di energia termica per la climatizzazione estiva ed invernale.

UNI/TR 10349-2 (2016). Riscaldamento e raffrescamento degli edifici – Dati climatici - Parte 2: Dati di progetto.

Voss K., Sartori I., Napolitano A., Geier S., Gonzalves H., Hall, M., Heiselberg, P., Widén, J., Candanedo, J., Musall, E., Karlsson, B. and Torcellini, P. (2010). *In*: Proceedings Eurosun Conference Graz, Austria, September 28-October 1, 2010, Load Matching and Grid Interaction of Net Zero Energy Buildings. doi: 10.18086/eurosun.2010.06.24

Vicente, R., Ferreira, T.M. and Da Silva, J.R.M. (2015). Supporting urban regeneration and building refurbishment. Strategies for building appraisal and inspection of old building stock in city centres. *Journal of Cultural Heritage*, 16(1): 1-14.

Wang, M. (2020). Dominator tree data flow cognitive analysis for green public building design. *The Journal of Supercomputing*, 76(2): 1268-1276.

Wang, Y., Zadeh, L.A., Widrow, B., Howard, N., Beaufays, F., Baciu, G. and Raskin, V. (2020). Abstract intelligence: Embodying and enabling cognitive systems by mathematical engineering. *In*: Cognitive Analytics: Concepts, Methodologies, Tools, and Applications, pp. 52-69, IGI Global.

Yang, Z. and Becerik-Gerber, B. (2014). The coupled effects of personalized occupancy profile based HVAC schedules and room reassignment on building energy use. *Energy and Buildings*, 78: 113-122.

Yau, Y., Wing Chau, K., Chi Wing Ho, D. and Kei Wong, S. (2008). An empirical study on the positive externality of building refurbishment. *International Journal of Housing Markets and Analysis*, 1(1): 19-32.

Yilmaz, Z.İ. (2019). Energy Efficiency and Rebound Effect for Household Gas Consumption: Evidence from Ankara. Doctoral dissertation. Middle East Technical University.

Yoo, W., Kim, H. and Shin, M. (2020). Stations-oriented indoor localization (SOIL): A BIM-based occupancy schedule modeling system. *Building and Environment*, 168: 106520.

Zero Carbon Hub (2014). Closing the gap: Design and as-built performance. *Evidence Review Report*, London: Zero Carbon Hub.

Zani, A., Tagliabue, L.C., Poli, T., Ciribini, A.L.C., De Angelis, E. and Manfren, M. (2016). Occupancy Profile Variation Analyzed through Generative Modelling to Control Building Energy Behavior. International High Performance Built Environment Conference – A Sustainable Built Environment Conference 2016 Series (SBE16). iHBE, 2016, 17-18 November, Sydney, Australia.

Index